Waveforms

A History of Early Oscillography

Waveforms

A History of
Early Oscillography

V J Phillips

Department of
Electrical and Electronic Engineering
University College
Swansea

Adam Hilger, Bristol

IOP Publishing Ltd 1987

British Library Cataloguing in Publication Data

Phillips, Vivian J.
  Wave forms : a history of early oscillography.
  1. Oscillograph — History
  I. Title
  621.3815′48   TK381

  ISBN 0-85274-274-6

Consultant Editor: **Professor A J Meadows**, Loughborough University

Published under the Adam Hilger imprint by IOP Publishing Ltd
Techno House, Redcliffe way, Bristol BS1 6NX, England

Typeset by Mathematical Composition Setters Ltd, Ivy Street, Salisbury.

Printed in Great Britain by J W Arrowsmith Ltd, Bristol

To my Parents,
Margaret M Phillips and the late Emlyn Phillips

Contents

Preface

The inability to 'see' what was happening in their circuits must have been a source of great frustration to the early experimenters in electrical science. The same was certainly true for the pioneers of acoustics and phonetics who needed a reliable means of rendering visible the vibrations of air and of sounding bodies. Many attempts were made to use comparatively crude mechanical devices such as rotating mirrors, sensitive flames and simple diaphragm recorders of various sorts to display these waveforms.

With the development of alternating current power systems during the latter part of the nineteenth century the need for a convenient, reliable 'wave-tracing' method (to use the jargon of the day) became ever more pressing, for if the operation of a generator supplying various types of consumer loads was to be properly understood and its performance improved, it was essential to make a detailed study of the various current and voltage waveforms. The arrival of the Joubert contact or 'point-to-point' principle in 1880 was a great step forward. Although the procedure was tedious and time consuming to use, at least it enabled progress to be made, and it remained the standard method of waveform delineation in electrical engineering for some twenty years.

The advent of the Blondel/Duddell mechanical oscillographs at the turn of the century was an event of paramount importance. At last there was a relatively quick and convenient method of waveform display, and the impact of this must have been tremendous. Something of the sense of euphoria can be discerned in the following passage describing the new oscillographs written by S M Kintner in the *Electric Journal* (vol III) of 1906.

> This very important piece of apparatus is being more generally used every day. It is no longer looked upon as an interesting instrument fit for laboratory service only. Engineers are rapidly finding uses for it that give them new insight into the working of their machines. Old theories are being checked or disproved.... The current and voltage variations resulting from the blowing of a fuse, the opening of a circuit breaker, the discharging of a lightning-arrester, the switching of transformers on high-tension lines and many other momentary conditions, can be studied with an oscillograph better than any

other way. It is also of equal importance in examining the waveforms of recurring phenomena. It is very frequently called upon in the diagnosis of serious cases of trouble and generally proves equal to the emergency.

The mechanical oscillographs, especially those produced by the Cambridge Scientific Instrument Company, were extremely successful instruments which remained in service for some thirty years or so. Indeed, they are occasionally used even today to fulfil certain special conditions—namely the provision of multichannel operation, often involving high voltages where good interchannel isolation is required.

In the present author's experience, today's physicist or engineer might, perhaps, have heard vaguely about the Duddell oscillograph, but probably has no clear idea of how it worked, and certainly has little knowledge of the other earlier methods employed to display waveforms. The purpose of this book is to document these various techniques which were used before the ubiquitous cathode ray oscilloscope swept them all into oblivion.

V J Phillips
Swansea, Wales, 1986

Acknowledgments

The gathering of information for a book such as this is a long and time-consuming process, involving much searching through the technical literature of the late nineteenth and early twentieth centuries. I am very grateful to the Director of the Science Museum, South Kensington for allowing me to spend some months there as Visiting Research Fellow, and also to the Librarian of the Science Museum Library and his staff for the facilities and help they have provided throughout. I am much indebted to Dr Brian Bowers and to Mr C N Brown for their considerable assistance in many different ways. When using facilities which are a long way from home, one has to try to use one's time to the maximum advantage and to achieve as much as possible on every visit, and in this connection I am very grateful indeed to the staff of the photocopying section of the Science Museum Library for their willing assistance in the production of copies of diagrams in the limited time available.

I am also indebted to Mr M A Cross of Maryland, USA for information concerning some of the Westinghouse instruments, and to Mr D H Gallagher of Pennsylvania for details of apparatus produced by the Leeds and Northrup Company. The assistance of Mrs F Bowers, Mr M F Lauer, and Dr and Mrs K J Gallagher in the translation of German papers is gratefully acknowledged.

My grateful thanks are also due to Mr J D S Morgan, Mrs J C Pugh, Mrs A M McGairl, Mr D R Gabriel and Mr W Pring of University College, Swansea for their help in the preparation of diagrams, and to Mrs F M McEwan, Miss L Forbes and Mrs J Thomas who assisted with the typing of the manuscript.

Most of the diagrams in this book have been reproduced from originals which appeared in the contemporary journals. After such a lapse of time, no formal question of copyright arises. However, copying from books is rather a different matter. Wherever it seemed possible that copyright might still subsist, every effort has been made to contact the publishers, or to trace the holders of such copyright. The help in this afforded by my colleague Mr G F Evans is acknowledged with thanks.

I would like to convey my thanks to all the publishers and institutions who have given permission for the reproduction of diagrams. They are listed below, and the individual references are given on the relevant figures. The kindness of Dr Jean Vergnes of Toulouse, who gave permission for the use of diagrams from Henri Bouasse's book, is also gratefully acknowledged.

Some of the material of Chapter 6 has already been published by the present author in a paper entitled 'Optical, chemical and capillary oscillographs' which appeared in The Proceedings of the Institution of Electrical Engineers **132** A (No 8, December 1985). The Institution therefore owns the copyright of this material, and permission to make use of it is acknowledged with thanks.

Finally I would like to express my gratitude to my wife, Diana, for her constant encouragement and support in this enterprise.

Permission to reproduce diagrams was granted by the following:

The Royal Society, London,
The Royal Institution, London,
The Royal Society of Edinburgh,
The Trustees of the Science Museum,
The British Library,
The Carnegie Institute, Washington,
Cambridge Instruments Ltd,
Macmillan and Co., London,
Macmillan Publishing Co., New York (Cassell & Co.),
McGraw-Hill Book Co., New York,
Longman Group Ltd, Essex,
Messieurs Gauthier-Villars, Paris,
J A Barth Co., Leipzig,
Electrical-Electronic Press (formerly Wireless Press),
Libraire Delagrave, Paris,

V J Phillips
Swansea, Wales, 1986

1

Introduction

The cathode ray oscilloscope is now a standard item of equipment in every electrical laboratory. The original crude display tube of Karl Ferdinand Braun and its associated circuitry have been refined and improved, particularly in the years since the Second World War, so that the modern instrument is convenient to use, has ample bandwidth for most purposes, and displays the required waveform on a graph with axes calibrated directly in volts and seconds. Periodic waveforms are presented as a steady trace by the use of sophisticated synchronising circuits; transient signals may be stored for extended examination either by digital methods or by phosphor-storage tubes. 'Instant' photography is also available for the preparation of permanent records, or else the stored digital information can be used to draw the waveform out on an $X-Y$ plotter.

We take these things so much for granted nowadays that it is very easy for us to forget the tremendous difficulties which beset the early experimenters who had to carry out their work without the aid of such a useful and versatile tool. An example drawn from the early nineteenth century will serve to illustrate this point.

In 1842, Professor Joseph Henry disclosed the results of some experiments he had been carrying out concerning the discharge of Leyden jars through coils of wire [1–4]. The Leyden jar was the form of capacitor in use in those days, and consisted of a glass vessel with inner and outer coatings of metal foil which formed the plates of the capacitor. Henry discovered that steel needles which were placed in the centres of the coils were left in a magnetised state after the discharge of the jar. He also noticed the most interesting, and rather puzzling, fact that sometimes the needle would be left magnetised in one direction, and sometimes in the other. It had generally been assumed that during discharge, the electricity flowed smoothly from one plate to the other, but using that assumption it was difficult to explain this apparently anomalous result. Further thought led him to suggest that the discharge actually occurred in an oscillatory manner, the electric charge surging backwards and forwards between the plates until the energy in the system was eventually dissipated.

These experiments began to cast some light on some earlier rather mysterious experimental results achieved by F Savary in 1826/7, when he was attempting to observe the magnetic field patterns around linear and helical wires through which batteries were being suddenly discharged [5]. He, too, observed a variation in the direction of magnetisation, and in his own words:

> The electric movement during the discharge is made up of a train of oscillations transmitted by the wire to the surroundings and soon damped by the resistances which rise rapidly with the absolute speed of the excited particles. All the phenomena lead to this hypothesis.

The suggestion that such oscillations could occur met with a certain amount of scepticism; after all, what could possibly account for such a thing? Today, we are quite used to dealing with the second-order differential equations which describe the flow of charge in circuits containing resistance, inductance and capacitance, and we know very well that under certain conditions, usually referred to as 'underdamping', the solution to the equations is oscillatory in form, but in the 1830s this analysis was quite unknown. In fact, it was not until 1853 that William Thomson (later to become Lord Kelvin) published the theoretical proof [6] and G R Kirchhoff produced a similar analysis some years later [7]. Even then, for many people, the writings of Thomson and Kirchhoff were merely theoretical abstractions which did not really mean very much. What they needed was concrete, visible proof which would settle once and for all, in a way which could be appreciated by everyone, the question of whether or not there were oscillations in the discharge.

We shall see in the next chapter how such proof was provided by Feddersen, using Wheatstone's technique of the rotating mirror, but the point which is brought home to us forcibly by these events is that in Henry's day, experimenters simply had no means of 'seeing' what was happening in their circuits. Even with the development of good ammeters and voltmeters, there was still no way of observing the detailed instant-by-instant variations of current and voltage—unless such variations were so slow that the inertia of the meters was of no consequence. How easily the question of Leyden jar oscillations could have been settled if some means of observing the current waveform, (or of 'waveform tracing' as it was often called) had been available to Henry and his contemporaries. Even as late as 1916, W R Cooper, referring to developments which had taken place at the very end of the nineteenth century, was moved to write [8,9]:

> ...before this time, oscillations and phase differences were familiar enough on paper, but to see curves trace themselves upon a screen and have the effects of capacity and inductance appear as if by magic must have reassured those who were weak in faith, whether students, or an older generation who fought shy of the theory of alternating currents.

Although a means of observing current variations would certainly have been very useful to the earlier experimenters, it was the development of alternating current power systems in the last two decades of the nineteenth century which provided the great impetus for the development of waveform-tracing instruments. If generators were to be designed properly, and if the distribution systems were to be operated at the best efficiency, it was clearly vital for both the designers and the operating engineers to be able to monitor what was going on. By that time, ordinary DC measuring instruments were in a reasonably well advanced state of development, and meters capable of measuring RMS values of alternating waveforms were also available, but there was still a need for a method of examining the waveforms themselves in detail. Figure 1.1 shows a timescale, extending from 1830 to the Great War of 1914–18, indicating some of the more important landmarks in the development of electrical science and engineering. Although the Swan and Edison filament lamps were invented in 1879, the somewhat earlier arc lamps continued to be used in a variety of forms for many years. As is well known, the arc lamp exhibits a very curious relationship between the voltage applied to it and the current which flows and, indeed, over part of its characteristic it actually has a negative resistance. Such a device played havoc with the waveforms on alternating current distribution systems, and a great deal of effort was expended in trying to analyse and understand this complicated situation. Both Blondel and Duddell, (and many others who also figure prominently in the story of the development of the oscillograph), were very much concerned with this problem, and a number of their papers describing oscillographic developments use arc lamp voltages and currents as illustrative examples. In Duddell's case, the paper in which he described his newly developed oscillograph was actually entitled 'Experiments on alternate current arcs'. Yet another complicating feature was the fact that the major components of power supply networks—generators and transformers—are constructed of iron and steel, which are magnetically saturated by excessively high fields, and can thus introduce all sorts of distortions and harmonics. Two quotations from the literature of the period will serve to illustrate the problems which the pioneers of AC power systems were facing:

Alternating currents, such as those provided by electromagnetic machines such as the 'Alliance', or the more recent and better types of Gramme and Siemens, have so far been studied very little; this can be explained by the difficulties which are encountered when one tries to apply ordinary instruments and methods to large currents which are changing polarity at 100 or 200 times per second. (Joubert 1880 [10,11])

For the most part, theories relating to alternating currents are based on the hypothesis, rarely verified, that the current is sinusoidal in form, whose period is controlled by the speed and number of poles of the alternator. In reality,

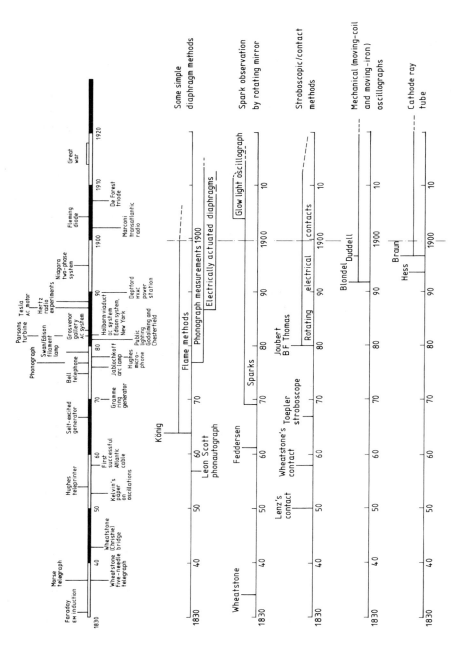

Figure 1.1 Chart showing some 'landmarks' in the development of electrical engineering from 1830–1914, together with an indication of the periods during which the various wave-tracing methods were in use.

one knows very well that even with a sinusoidal e.m.f. the introduction into the circuit of other equipment produces deformation of the curve. (Armagnat 1897 [12])

Many waveform display devices were developed with the investigation of these alternating current phenomena specifically in mind.

Although the needs of the electrical engineer came to dominate the work on waveform display, it was another field of activity which really provided the initial impetus, namely that of acoustics with its sister science of phonetics. Much of the early effort in this field involved the use of what might be termed simple, direct methods. Thus the vibrations of tuning forks or strings were recorded by attaching a bristle to the vibrating member, and allowing its free end to press lightly against a moving-glass plate, previously smoked over a flame. The movements were thus inscribed and recorded. When it was required to investigate the air pressure variations produced by the human voice or by a musical instrument, the sounds were allowed to impinge upon a diaphragm, the resulting vibrations of which were again recorded in the same direct manner.

One of the most important and far reaching of all the nineteenth century's innovations was the introduction by Wheatstone in 1834 of the principle of observation through a rotating mirror. The general idea may be simply expressed as follows. If a steady light or a stationary body is observed through a rotating mirror, a simple band of light or a straight line representing the edge of the body is seen. However, a flickering light appears as a band with light and dark striations, and the edge of a vibrating body as a wavy line. Suitably recorded, either by simple free-hand drawing or by photography, the resulting trace enables one to deduce the instant by instant movement of the object under observation. As previously mentioned, it was this principle which enabled Feddersen to furnish proof of the Leyden jar oscillations, for the spark resulting from an oscillatory current varies in luminous intensity, and these variations are revealed by the movement of the mirror.

The importance of the rotating mirror as a general technique must be stressed, for until the development of the scanned cathode ray beam, it was the mirror which produced the time scan in all the mechanical oscillographs. Before that time, it was used in conjunction with the simple diaphragm devices of the phonetician to render the vibrations more easily visible. The German experimenter Rudolf König, for example, allowed the vibration of the diaphragm to vary the pressure of the gas supply to a sensitive flame, which oscillated up and down in accordance with the spoken sounds. When viewed through a rotating mirror, the resulting bright/dark boundary line appeared as a waveform—of sorts.

The arrival of the Edison phonograph in 1879 was an important milestone in the development of simple diaphragm devices, for not only was it a way of preserving sounds for posterity, it was also a method of recording the movement of the diaphragm in a permanent way. The recording could then be measured microscopically, and the readings could be plotted to provide a graph of displacement against time.

Although the needs of the electrical engineer and the acoustician arose from quite different causes, those needs actually coalesced as time went on. It was a relatively simple matter to substitute a coil-actuated, telephone type of diaphragm in the flame apparatus or the phonograph so that the electrical current waveform could be observed. On the other hand, after the invention of the microphone by Bell, Hughes and others in the 1870s, air vibrations could easily be converted into electrical variations, so that most of the apparatus introduced by the engineers for their own purposes could also be used by the acoustic experimenters.

The first method of tracing the waveforms of electrical quantities was by observation of the spark. This was actually not at all satisfactory for detailed waveform examination since the spark was a very volatile object, the brightness of which bore only a rough relationship to the current flowing. An attempt was made to refine this method early in the twentieth century by making the discharge occur in a glow tube. When properly constructed, the length of the glow discharge along the electrode in the tube could be made proportional to the current flowing. When viewed through the usual rotating mirror, a much better representation of the waveform was provided than that obtained with the simple spark. However, this method, although ingenious, did not find much practical application because the traces produced by photography were really very fuzzy and ill-defined.

It was the introduction of Joubert's contact method in 1880 which represented the first real breakthrough made by the electrical engineers. This method, which was actually a stroboscopic contact mounted on a shaft rotating synchronously with the waveform under observation, was extremely slow and tedious to use since, in the simple form of the apparatus, the contact required resetting for each point measured on the curve. However, under the appropriate circumstances, it was a method capable of providing an accurate and detailed picture of the alternating current or voltage wave, and it remained the standard procedure for two or three decades. Many schemes were suggested to enable the process of measurement to be speeded up and to render it semiautomatic or wholly automatic in operation. The names of Professor Rosa in the United States and Monsieur Hospitalier in France were prominent in this field. Interestingly, the contact method also sprung up in a completely different field of endeavour, namely electrophysiology. Here they were usually referred to as 'rheotomes', this being a word coined by Wheatstone to denote a device for periodically interrupting a current. Although they were used here in a slightly different way in order to investigate the electrical signals associated with muscular activity, they were still fundamentally contacts taking samples of the waveforms, and as such they were closely related to the Joubert contacts. The chart shown in figure 1.2 attempts to relate the different methods of observation to the different fields of endeavour in which they were developed, although it must be stressed that this is a very simplified presentation.

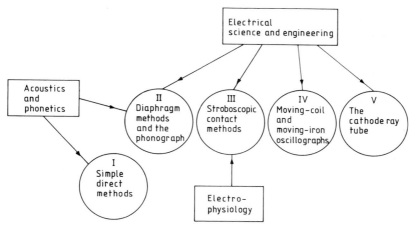

Figure 1.2 Chart showing areas of research for which the various waveform-tracing methods were developed. The methods are numbered in approximately chronological order although, of course, there was some overlap.

The contact method was eventually displaced by the mechanical oscillographs which arrived during the last decade of the century. To André Blondel is due the credit for the initial invention of both the single-loop moving coil and the moving-iron versions. At first Blondel concentrated his efforts on the moving-iron version, but later in the decade he returned to the moving-coil idea, and developed several successful versions of it. Meanwhile, the single-loop principle had been taken up by William Duddell in Great Britain and, in collaboration with the Cambridge Scientific Instrument Co., he produced a series of well thought out and competently engineered oscillographs which stood the test of time and remained in use until well after the Second World War. Indeed, they are occasionally employed even today for certain specialised applications. The Joubert contact and the mechanical oscillographs between them can be said to have seen electrical engineering through the most innovative phase of the growth of power generation and distribution at the end of the nineteenth century.

As might have been expected of the resourceful Victorians and Edwardians, there were several other very ingenious (but often highly impracticable) methods, outside the main lines of development, invented for the purpose of waveform display. None of these succeeded to any great extent; nevertheless they are interesting in their own right and they certainly illustrate the inventiveness of our predecessors, and the way in which they were prepared to try anything, however improbable, which held out even a slender promise of achieving that most desirable end, namely rendering visible what was happening in electrical circuits, or in the vibrating air.

Of course, it was the cathode ray tube which eventually conquered all and became the universal method of waveform observation. The original germ of the idea of using cathode rays for waveform tracing seems to have been due to Hess, a Frenchman, but it was undoubtedly Braun, the German scientist, who took the first positive steps. Since it is now so ubiquitous an instrument, and since it clearly has many advantages compared with the methods it displaced, it would be natural to suppose that when the cathode ray tube was introduced it was immediately hailed as the answer to all the problems. However this was not the case and, in fact, it really took quite a long time to reach full perfection in the state in which we know it today. When compared with the speed of development of the mechanical oscillograph of Duddell, for example, it really was a rather slow process. Perhaps the gain in ease of operation to be achieved in going from the mechanical oscillographs to the cathode ray instrument was less than that in going from the Joubert contact to the oscillograph, and so the 'market forces' urging on the work would accordingly have been that much less.

The aim of the present book is to tell the story of waveform observation from the time of Wheatstone up to the introduction of the cathode ray tube. Many of the earlier methods are now all but forgotten. In the present author's experience, today's engineer has usually heard vaguely about the Duddell oscillograph (but has a very hazy idea of how it worked), but has no knowledge at all about flames, phonographs, contacts etc. and all those other methods which have long since fallen into disuse. The chapters of the book will follow more or less the historical sequence outlined above.

Chapter 2 will deal firstly with the rotating-mirror and spark methods, and with their ultimate manifestation as the glow light oscillographs. It will then go on to the simple, direct methods; the vibrating diaphragms and the various methods employed to record those vibrations—by sensitive flames, smoked drums and the phonograph.

Chapter 3 covers the rotating contacts, starting with the simple primitive system, and proceeding to the development of semiautomatic recording methods, and fully automatic instruments such as Hospitalier's 'ondograph'. The chapter finishes with an account of the contact methods developed by the electrophysiologists.

The next two chapters deal with the mechanical oscillographs, Chapter 4 concentrating on the moving-coil principle, and Chapter 5 on the moving-iron versions and on the so-called 'compensated' instruments which used what were, effectively, electrical weighting or predistortion networks to allow for the inherent deficiencies of the system, and to improve the frequency response and the fidelity of delineation of the waveforms.

Chapter 6 will be concerned with all those miscellaneous ideas which were tried out as alternatives to the main lines of approach to the problem; i.e., with all those isolated individual methods which do not fit in conveniently with the main lines of development described in the other chapters.

Finally, to round off the story, Chapter 7 will deal very briefly with the beginnings of the cathode ray tube and with its subsequent development.

The chart shown in figure 1.1 shows, in a simplified way, the periods of time during which these various techniques were of practical importance. It will be appreciated that whilst it is usually fairly easy to state with precision the date of the initial introduction of a method, since this can be ascertained from the date of the original published article or paper describing it, it is less easy to say when a given method ceased to be used. We all know from our own experience that if we have an instrument in our laboratories, we continue to use it until such time as we can afford something better or more up to date!

References

[1] Report 1842 *Proc. Am. Phil. Soc.* **2** 193–6
[2] Taylor W B 1880 in *A Memorial of Joseph Henry* (Washington:US Govt Printing Office) (Smithsonian Publications) p 255
[3] Crowther J C 1937 *Famous American Men of Science* (London: Secker and Warburg) p 193 (also 1946 (London: Pelican) pp 159 ff)
[4] Jordan D S 1910 *Leading American Men of Science* (New York: Holt) pp 127ff
[5] Savary F 1827 *Ann. Chimie Phys.* **34** 5–57
[6] Thomson W 1853 *Phil. Mag.* **5** (series 4) 395–405
[7] Kirchhoff G 1864 *Poggendorff's Ann.* **121** 551–66
[8] Cooper W R 1916 *Electrician* **77** pp 793–6
[9] Cooper W R *Forty Years of Electrical Progress* (Reprint of papers in *Electrician* **77**: published by Electrician) (see Introduction, p 7)
[10] Joubert J 1880 *J. Physique* **9** 297–303
[11] Report 1880 *Electrician* **5** 151–2
[12] Armagnat H 1897 *L'Éclairage Électrique* **12** 346–53

2

Mirrors, Flames and Membranes

Summary

In the first section of this chapter, after an account of the introduction of the rotating-mirror principle by Wheatstone, the apparatus incorporating it which was used by Feddersen to furnish a proof of the oscillations in Leyden jars, will be discussed. More refined versions of this apparatus were produced over a surprisingly long period—well beyond the turn of the century in fact—and some of these improvements will then be covered. The glow light oscillographs which follow represent the ultimate development of the method and its extension to true waveform tracing.

Attention in the second section will be focussed on the efforts of experimenters in the field of acoustics to record sound waves. König's use of the sensitive flame to render visible the vibrations of diaphragms or membranes will be described, together with later variants of it. The severe distortions inherent in this technique meant that a very crude version of the required waveform was produced. Some of the simpler, more direct methods used to record the movement of vibrating objects with rather better fidelity will then be presented.

The final section will deal with the introduction of the phonograph, and the use that was made of it for acoustic measurement, particularly for the delineation of the waveforms of speech and music.

Since many of the techniques dealt with in this chapter were basically methods of recording the movement of a diaphragm, it was relatively easy to substitute the diaphragm of a telephone ear piece for the simple acoustic membrane, and hence to make measurements of electrical waveforms such as those associated with AC supply systems. These electrical applications will be described where appropriate throughout the chapter.

2.1 Observations of Sparks Through Rotating Mirrors

The first published reference to the rotating mirror seems to have been in one of Wheatstone's papers, published in 1834, entitled 'An account of some experiments to measure the velocity of electricity and the duration of electric light' [1,2]. The form of rotating mirror he used can be seen in figure 2.1. It was mounted on a vertical spindle which rotated at a speed of 50 revolutions per second, this speed being such that 'the reflected image of a luminous point passed over one half of a degree in one 72,000th of a second'. The introductory section of the paper refers to some previous work which he had carried out 'relating to the oscillatory motions of sonorous bodies', but those experiments do not appear to have been written up by him in the form of a paper presented to a learned journal. In the 1834 paper however, he does give some brief details:

> Vibrating bodies afford many instances for investigation; one among these is perhaps worthy to be mentioned. A flame of hydrogen gas burning in the open air presents a continuous circle in the mirror; but while producing a sound in a glass tube, regular intermissions of intensity are observed which present a chain-like appearance and indicate alternate contractions and dilations of the flame corresponding with the sonorous vibrations of the column of air.

The mirror could also be tilted about its horizontal axis so that the light from the object under investigation would be displayed as a circular band on the ceiling. Thus 'the image of the sun is converted into a magnificent fiery belt'. Any oscillation in the light source would, of course, show up clearly as discontinuities or dark segments in the band.

Figure 2.1 Wheatstone's rotating mirror [1]. Courtesy: Royal Society.

In Wheatstone's first experiment, he attempted to measure the speed with which an electric spark jumped across a gap. The general idea was that if the charge took a finite time to move across the gap, the image viewed in the rotating mirror would appear as a sloping line, each part of the spark becoming luminous at slightly different instants of time. The experiment yielded a negative result, the sparks 'appearing perfectly unaltered and precisely as they would have done had they been reflected from the mirror while at rest'. This indicated that the charge travelled across the gap instantaneously—or at any rate, faster than could be measured with a mirror revolving at 50 revolutions per second.

The second experiment which he performed was an attempt to measure the speed of propagation of an electric disturbance along a wire. Half a mile of insulated wire was suspended from the balustrade of the gallery in Adelaide Street in London in a zig-zag manner. There were spark gaps at the ends of the wire and also at the centre of its length. The three spark gaps were mounted close together on a wooden baseboard as in figure 2.2, and were observed through an improved rotating-mirror system. The mirror (E), mounted on a horizontal axle, can be seen in the detailed diagram of figure 2.3 and the disposition of the entire apparatus is shown in figure 2.4. The Leyden jar capacitor was charged up through the connection N and the charge was allowed to pass to the balls and the ends of the zig-zag wire through a 'switch' ball Q mounted on the rotating shaft. It was so arranged that the sparks would occur when the mirror was in a suitable angular position for viewing. To make this switching action rather more precise, the trigger spark had to pass from the rotating ball to the stationary one through a fine horizontal slot cut in a screen of mica (S).

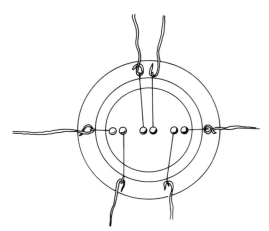

Figure 2.2   Wheatstone's three spark gaps, mounted side by side on a base board [1]. Courtesy: Royal Society.

The rotating shaft was driven by hand through a gear train and pulley system. Various methods were used to determine the precise speed of rotation of the shaft, including the mounting of a compressed-air siren on the shaft. This proved to be rather difficult to use, and he eventually measured the shaft speed by allowing the ball Q to beat against a piece of paper as it rotated, producing a musical note. It is reported in Wheatstone's paper that the note being of pitch $G^{\#4}$ meant that the shaft was rotating at 800 revolutions per second.

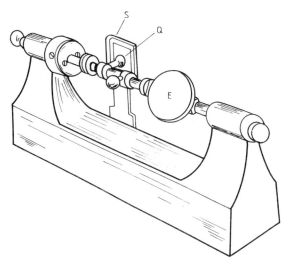

Figure 2.3 Rotating mirror (E) with trigger spark gap (QS) [1]. Courtesy: Royal Society.

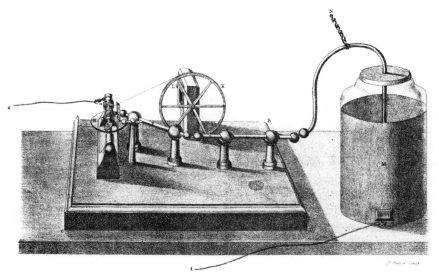

Figure 2.4 Wheatstone's complete apparatus [1]. (Figure 2.3 is an enlarged view of the section at the left.) Courtesy: Royal Society.

Wheatstone drew several conclusions from his measurements. The shaft speed of 800 revolutions per second which he eventually achieved was much greater than the 50 revolutions per second which he had used in the first experiments (mentioned above), and this enabled him to correct his first result, and to estimate that the duration of each spark was finite, but 'less than one millionth part of a second'.

Wheatstone observed that the two sparks at the ends of the line occurred simultaneously, but the one at the centre took place a fraction later. As far as the determination of the velocity of propagation of electricity along the line was concerned, his conclusion from a measurement of the time delay was that 'the velocity of electricity through a copper wire exceeds that of light through the planetary space'. He stressed, however, the inherent limitations of his measurements, stating that he was 'not prepared to state results with numerical accuracy', and he made it clear that further, more refined experiments would be needed to confirm this figure. Wheatstone's anomalous result can be ascribed to the fact that his transmission line was not laid out in one long length, but was constructed in a zig-zag fashion so that there was mutual coupling between the sections of the wire.

He also drew one other interesting conclusion concerning the nature of electricity from his results. For a number of years, argument had raged as to whether electricity flowing in a wire consisted of a single type of electric fluid passing from one end to the other, or whether it was actually two different kinds of fluids travelling simultaneously in opposite directions. In 1733, Charles-François du Fay (or Dufay), Superintendent of the Gardens of the King of France, had postulated the two-fluid theory. He had carried out some experiments on the electrification of various substances by friction (notably resin and glass), and by consideration of the fact that sometimes charged bodies experienced mutual attraction, sometimes repulsion, he concluded that there were two kinds of electricity which he called 'vitreous' and 'resinous'. These ideas were further developed by the Abbé Nollet who introduced the idea of the different fluids flowing simultaneously in two directions.

Around 1747, Benjamin Franklin proposed instead a one-fluid theory. This single electric fluid was supposed to be present in all matter, and the various attractions and repulsions could be explained in terms of excesses or deficiencies of the fluid in the electrified bodies. Wheatstone saw in his results a proof of the correctness of the two-fluid theory, since the sparks at the ends of the wire had occurred simultaneously, In his own words '. . .on the hypothesis of an actual transfer of a fluid from one end of the wire to the other a difference of time between the two sparks at the opposite extremities might be expected to be observed. . .on the supposition of the transfer of two fluids in opposite directions, the extreme sparks would be simultaneous, but the middle spark later'. This whole controversy between one- and two-fluid electricity rumbled on unresolved for many years until it was overtaken by modern electronic theory at the end of the century.

From the point of view of the present book on the subject of waveform delineation, Wheatstone's introduction of the principle of the observation of rapidly changing phenomena through a rotating mirror is of the greatest importance. This was a technique which thereafter was to be used very frequently by many people. Wheatstone's contribution was widely acknowledged and, in fact, the technique was often referred to in nineteenth century books and papers as simply 'Wheatstone's mirror'. Wheatstone also proposed to improve his apparatus by replacing the simple mirror by a six-sided mirror drum mounted on the shaft. As he pointed out 'If the light be transient, we shall have six times the number of chances of observing its reflection than if one reflecting surface only were employed.' This mirror-drum principle was to become one of the most commonly used methods of observing oscillating objects, and was to be widely employed in the mechanical oscillographs which will be discussed in Chapters 4 and 5.

Let us now return to the question of oscillations in the discharge of the Leyden jar. In the theoretical paper of 1853, previously mentioned, [3] entitled 'On transients in electric circuits' William Thomson derived the well known equation governing the flow of charge q.

$$\frac{d^2q}{dt^2} + \frac{K}{A}\frac{dq}{dt} + \frac{1}{AC}q = 0$$

K representing the resistance, A the inductance and C the capacitance in the circuit. Having explored all the various implications of this expression, he went on to say:

> It is probable that many remarkable phenomena which have been observed with electric discharges are due to the oscillatory character we have thus found to be possible when the condition $C < 4A/K^2$ is fulfilled: rapidly succeeding flashes of lightning might thus be explained. A corresponding phenomenon might be produced artificially on a small scale by discharging a Leyden phial across a very small space of air.... Should it be impossible on account of the too great rapidity of successive flashes for the unaided eye to distinguish them, Wheatstone's method of a revolving mirror might be employed which might show the spark as several points or lines of light separated by dark intervals instead of an unbroken line.

It was indeed a rotating mirror which furnished positive practical proof of oscillation in the hands of B W Feddersen in 1861 [4], and Thomson's perceptive remark, in fact, predicted the introduction of a measuring technique which remained an important tool for the investigation of the waveforms of varying phenomena until the end of the nineteenth century and beyond.

Feddersen's experimental arrangement can be seen in figure 2.5. A concave mirror mounted on a vertical rotating shaft driven by a weight through the gear train c can be seen at the left-hand side of the diagram. The Leyden jar on the right side was charged up and was then allowed to discharge through the spark gaps p and p', the moving image of the sparks being

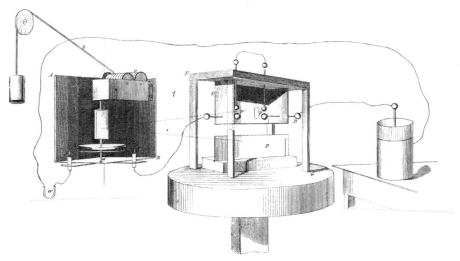

Figure 2.5   Feddersen's rotating-mirror apparatus ([4], Table VII on p 512).

focused onto the photographic plate P. So that the spark would occur when the mirror was in the correct position, the apparatus was provided with a trigger circuit consisting of a needle mounted on the mirror shaft which only permitted the spark to occur when it was in line with the two small spark-balls mounted at the bottom of the mirror apparatus. The U-shaped tube (W) was filled with mercury, thus including some resistance in the circuit. A single, unidirectional flow of charge during the duration of the spark would have resulted in a continuous line being recorded on the photographic plate. Many of the photographs showed, however, that the emission of light from the spark was in the form of a series of pulses, each one representing a separate and distinct surge of charge between the plates of the Leyden jar. The actual photograph published in Feddersen's paper is reproduced in figure 2.6(*a*). Thus the oscillating nature of the discharge was proved practically; plain for all to see and appreciate. As time went by he improved his technique and in another paper which appeared a year later, in 1862, he presented some really excellent photographs such as the one shown in figure 2.6(*b*) [5].

Berend Wilhelm Feddersen was awarded his Doctorate from the University of Kiel for his work and, in the words of *The Dictionary of Scientific Biography* (Scribner NY) 'Thomson's public acknowledegment of his debt to Feddersen brought the latter worldwide renown'. Feddersen was also one of the moving spirits behind the production of *Poggendorff's Biographical Handbook*, a work which is a most valuable source of information about the scientific workers of the time. Unfortunately, the large sum of money which he left for the continuation of the work was effectively rendered worthless by the great German inflation of the 1920s.

(a)

20

Figure 2.6 (*a*) Photograph of spark from Feddersen's 1861 paper ([4], Table VII on p 512)—proof of the oscillatory nature of the discharge of the Leyden jar. (*b*) Improved photograph from 1862 paper ([5], Table I on p 192).

These classic experiments were widely repeated and refined by other workers—by Ricardo Felice, Professor of Physics at Pisa in 1864 [6] for example. Ludwig V Lorenz, a Danish civil engineer of wide interests (credited, among other things, with the first accurate estimate of what is now known as Avogadro's number) also performed oscillatory spark experiments in 1879 [7]. The resolution possible in Feddersen's experiment was limited by the relatively low speed of rotation of the mirror shaft (about 100 revolutions per second) and much effort was expended in trying to increase this. O N Rood in 1869 [8], using a set of apparatus very similar to that of Feddersen (figure 2.7), succeeded in raising the speed to 300 revolutions per second. He used a plane mirror with a separate lens for focus, finding that this gave greater flexibility in positioning the various parts of the system. The spark was set off at the correct instant by a brass wire (W) mounted on the mirror shaft, and the light was brought to a focus on the ground-glass screen G after reflection from the plane mirror M. The image could be viewed through an eye-piece lens, and to assist in making measurements the image of a scale (C), illuminated by a small lamp, could also be thrown upon the screen. Rood seems to have been more concerned with the physical appearance of the spark rather than the purely electrical phenomena in the circuit, but he was able to make quite detailed measurements of the durations of the various parts of the discharge. In a follow-up paper written two years later [9] he claimed that further refinement of his equipment was enabling him to make measurements of time intervals of the order '28 billionths of a second, and possibly up to 10 billionths of a second'. Since he was writing for an American journal we must assume that this means 10×10^{-9}, or 0.01 μs.

The accuracy of the measurement depended, of course, on the accuracy with which the angular velocity of the mirror was known. A drum mounted on one of the shafts of the gearing wound up a band of paper, and a seconds pendulum was arranged to make a mark upon it once in each swing. The mirror speed could then be calculated from a measurement of the distance between marks and a knowledge of the gear ratios. He also made marks on the paper by hand, timing them by means of a watch with a seconds hand. In view of this, we should perhaps treat the accuracy claims with some scepticism.

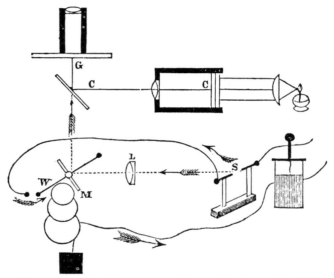

Figure 2.7   Rood's apparatus for spark observation [8].

Another prominent experimenter in this field was Professor J Trowbridge of the Boston Technological Institute and Harvard University, collaborating with W C Sabine [10–12]. Like most of the other people who performed this type of experiment, their aim was to increase the accuracy of measurement and to make detailed investigations of the effects of the circuit parameters on the frequency of the oscillation. Their particular concern in this case was with the influence of the dielectric material of the capacitor.

In a paper published in 1890 they described the result obtained with a large rotating plane mirror driven by a gas engine at a rate of 50 revolutions per second. A brush contact mounted on the shaft touched a stationary piece of brass and caused the discharge to occur at the appropriate moment. A large concave mirror with a radius of curvature of 313 cm and an aperture of 16.5 cm reflected the image of the spark onto the plane mirror and thence onto a photographic plate 260 cm away. The adoption of a plane rotating mirror with a concave stationary mirror enabled them to place the photographic plate at a distance and hence to obtain good magnification and a well spread out trace without the use of an excessively high rotational speed. Even so, the experiment was not without its hazards:

> Great care was taken to balance the mirror. Its large size and weight made it important on account of the danger of it flying apart that it should revolve with uniformity. The axis of the mirror was placed horizontally. This precaution proved to be a wise one, for twice in the course of the many runs which were made, the mirror flew into pieces; the excursion of the fragments however were confined to the vertical plane. This liability to accident is perhaps inherent in a method which employs a large plane mirror.

In Trowbridge's apparatus, the speed of the mirror was determined in the following way: a brass cylinder covered with lamp-blacked paper was mounted on the mirror shaft, and alongside it was a stylus moving along a stationary shaft parallel to the drum and so arranged that it could be slid along by hand, thereby drawing a spiral line on the paper. The output winding of a Ruhmkorff spark coil was connected between the stylus and the cylinder. The primary make-and-break mechanism of the coil was operated by the swinging of a seconds pendulum. When the stylus had started to move along the rod, it was so arranged that the occurrence of a spark from the coil released another pendulum which had hitherto been held up by an electromagnet. This acted as a delaying mechanism in conjunction with the trigger contact so that the Leyden jar was discharged when the mirror was suitably placed. Another spark from the stylus to the drum marked the end of a period of one second, the two sparks having left well defined spots on the lamp-blacked paper. A spiral of about fifty turns was produced by the stylus, and it was claimed by the authors that it was possible to measure time to an accuracy of 1/500th of a second in this way, a claim which seems quite modest when one realises that this time represented a drum revolution of 36° or so.

Reading between the lines of this paper, one gets the impression that the various contact mechanisms, particularly that of the rotating brush, gave a fair amount of trouble and in a later paper entitled 'Dampening (sic) of electrical oscillations on iron wires' a new and more reliable contact mechanism was described [13]. A sharp cutting tool, of the sort used on a lathe, was mounted on a disc on the rotating shaft and contact to it was made through a slip ring. A lever carrying a pad of soft type-metal† was moved forward when it was desired to trigger the spark so that the tool cut a groove into the metal, thereby ensuring a good solid metal-to-metal contact and positive operation. It was claimed that it was further improved by use of 'a wax of peculiar composition' coated onto the surface of the type-metal.

† An alloy of lead, antimony and tin.

Trowbridge carried out numerous experiments on low-voltage sparks and high-voltage sparks in circuits of various types, [14,15] and he published numerous photographs taken with the aid of the rotating mirror, that of figure 2.8 being an excellent example where the discontinuous and repetitive nature of the discharge can be clearly seen.

Figure 2.8   Spark photograph published by Trowbridge in 1894 [14].

The type of rotating mirror developed by Louis Décombe at Paris in 1898 [16] with the aim of studying oscillations up to 5 MHz in frequency was rather safer to use than Trowbridge's large rotating mirror. As he very sensibly remarked 'one cannot increase indefinitely the speed of the mirror. The greatest value which it can obtain is determined by the mechanical resistance to rupture of the rotating parts. In particular, as a matter of precaution one gives to ω a value ω_1 sensibly below that critical value'.

His mirror system, seen in figure 2.9, was concave, made of 3 mm thick glass and set in a solid steel mount which was spun at speeds up to 500 revolutions per second, driven by a belt from an electric motor. With this apparatus he was able to count up to fourteen separate sparks in a single discharge of his capacitor and was also able to show that a single frequency of sparking, which depended on the value of the capacitor, was produced. Décombe was, incidentally, a pupil of the well known scientist Gabriel Lippmann whose capillary electrometer will be described in Chapter 6.

Figure 2.9 Décombe's sturdily constructed mirror system [16].

One of the most elaborate pieces of equipment involving a rotating mirror was that developed by A Battelli and L Magri, and described in a paper with the title 'On oscillatory discharge' published in 1903 [17]. Angelo Battelli was Professor of Physics, first in Padua, and later in Pisa. One is perhaps slightly surprised to discover that even at this relatively late date they were still investigating 'the influences capable of modifying, in the case of discharge, the period of oscillation'. They were still trying to provide proof of some of the results of Thomson's theoretical predictions. In some preliminary experiments, their mirror shaft was driven by a clockwork motor, but this proving inadequate, they devised instead the turbine mechanism shown in figure 2.10. The steel mirror S on the shaft A was rotated by turbine C as a result of high-pressure steam or compressed air admitted to the chamber P. The published paper is not specific as to the top speed produced, but simply says that it was well in excess of 450 revolutions per second. The speed was measured by allowing a fine bristle mounted on the shaft to make marks on a smoked drum once in every revolution. At the same time, an electrically maintained tuning fork equipped with another bristle made a second trace on the drum, and by comparison of the traces the angular speed could be found. The complete apparatus was fairly complicated as can be seen from figure 2.11. The smoked recording drum was

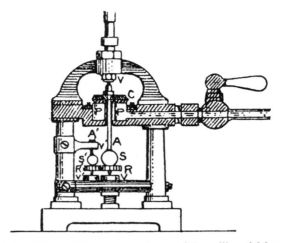

Figure 2.10   The turbine-driven mirror of Battelli and Magri [17]

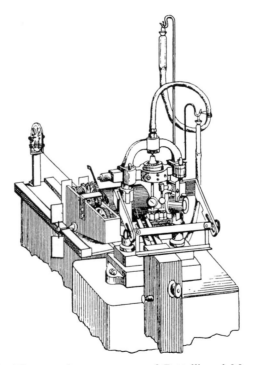

Figure 2.11   The complete apparatus of Battelli and Magri showing weight-driven recording drum and tuning fork for calibration of the trace [17].

turned by a weight acting though a gear train and regulated by an air vane; this can be seen on the left of the apparatus. The tuning fork is at the front. The optical system was quite straightforward; the image of the spark (which occurred inside a closed box) was formed on a photographic plate by a lens, and an ordinary pneumatic shutter was used to control the exposure.

Another worker who used the rotating-mirror principle to show the oscillatory nature of the capacitor discharge was E W Marchant [18], who carried out some experiments at Lord Blythswood's laboratory in Renfrew. Sir Archibald Campbell retired from the army after having been wounded at Sevastapol in the Crimean War. He served as an MP for many years, and was created Baron Blythswood in 1892. He was a devoted amateur of science and built himself a superb laboratory at Blythswood House. He is remembered chiefly for his work on optical diffraction gratings, although he also concerned himself with many other spheres of scientific investigation. He shared his laboratory with many eminent scientists of the day, and allowed them to make use of his excellent facilities for their work.

The Swiss-born physicist and engineer, L Zehnder [19–21] also used the rotating-mirror principle and he arranged for the images of the sparks to be projected onto a screen for purposes of demonstration to an audience. Zehnder's plane mirror was mounted in a depression cut into the side of a wood or ebonite cylinder (figure 2.12), and the triggering contacts (dd) were placed so as to fire the spark at the moment when the reflection in the mirror was directed at the projecting lens L.

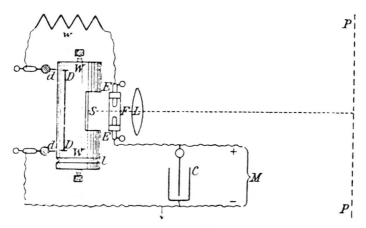

Figure 2.12 Zehnder's rotating mirror [19].

Optical methods which may be considered as variants on the rotating mirror were also employed. These included the rotating lens-disc devised by C V Boys in 1890 [22,23]. His experiments were slightly more general in nature than many of the others, and as well as examining sparks, he also

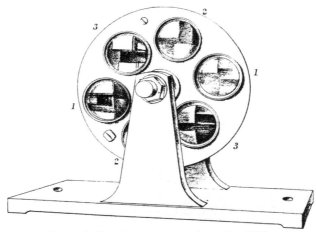

Figure 2.13 Boys' rotating lens-disc [23].

used the apparatus to view moving objects, such as falling water drops. In his arrangement, shown in figure 2.13, six lenses are mounted on the disc in such a way that each lens casts an image of the spark onto a separate circular track of a photographic plate, producing the type of trace seen in figure 2.14. The speed of rotation of the disc does not seem to have been very great; 'over 100 turns per second' is mentioned in the published paper. The intermittent nature of the spark is clearly apparent from the photographic traces produced. The fact that the intensity of succeeding sparks gradually diminishes due to the damping is also demonstrated. Arthur Schuster at Owens College, Manchester and his collaborator the Polish scientist, G Hemsalech [24] reversed this procedure in that they used a stationary lens projecting the image onto a rotating film disc, a speed of 120 revolutions per second being achieved. They were mainly interested in studying the movement of vapourised metal around the spark electrodes.

Let us now pause to consider precisely what all these experiments had actually achieved. All the pieces of apparatus described so far produced a

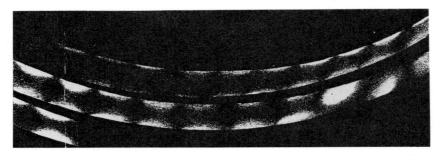

Figure 2.14 Photographic traces produced by the disc of figure 2.13 [23].

crude record of the current flowing in the circuit, made visible by the incandescence of a spark jumping across an air gap. The traces produced were in the form of an intensity versus time display, the intensity of the trace being a measure of the light emitted by the spark at any time, and hence some sort of indication of the magnitude of the current flowing. This was useful as far as it went, and certainly permitted experimental confirmation of Thomson's predictions to a moderate degree of accuracy. However, it was not really the type of record which would have been most useful. What was wanted ideally was a true graph of voltage or current plotted as a function of time. An interesting attempt was made to extend the spark and mirror principle to provide such a display. This took advantage of a phenomenon which had been investigated in 1902 by N Hehl [25,26] and by H A Wilson [27]; namely that in a Geissler gas discharge tube the extent of the glow on the negative electrode depends on the current flowing, and that the current density is independent of the area of the glow. Two people independently developed this phenomenon in the years 1904–1906 into what was known as the 'glow light oscillograph' (glimmlicht-oszillograph in the original German) or the 'cathode-glow oscillograph'; they were Ernst Gehrke [28] and Ernst Ruhmer [29,30], both of Berlin, Gehrke at the Imperial Physical and Technical Establishment at Charlottenburg; Ruhmer at his privately owned laboratory. A chance remark made by the Editor of the *Electrician* [31] referring to 'Ruhmer's ingenious oscilloscope' brought forth a letter from Dr Gehrke [32] claiming priority in the invention and quoting chapter and verse to prove it. (The Editor's early use of the word oscilloscope† may be noted here. This did not really gain currency until the 1930s or later.) In fairness to Ruhmer himself it must be said that the very first sentence of his paper makes reference to Gehrke's work. The active element in these glow light oscillographs was usually referred to as 'Gehrke's tube' and it is shown in its simplest form in figure 2.15. It consisted of a glass envelope containing nitrogen at a pressure of 6 to 8 mm of mercury, furnished with two long wire electrodes separated by a gap at the centre of the tube. When a DC voltage of sufficient magnitude is applied, the tube will strike and the gas will glow. As the voltage is increased, the cathode glow will spread further and further along the electrode, the length of the glowing column being proportional to the applied voltage. A change of polarity of the applied voltage results in transfer of the glow to the other electrode. If an alternating potential is applied to the tube, then to the naked eye equal lengths of glow will be seen on the two electrodes. When viewed through a rotating mirror however, the alternating waveform will be seen. Ruhmer's apparatus can be seen in figure 2.16, the tube being vertical in the centre, and the rotating mirror in the form of a mirror drum on a motor shaft appearing at the right. The discharge tube (figure 2.17) is

† From skopein—to see.

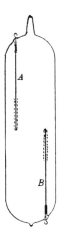

Figure 2.15   Gehrke's gas discharge tube [28].

of slightly different construction from that of Gehrke, consisting of two cylindrical electrodes mounted in line and separated at the centre by a thin mica spacing plate with a small hole at its centre. The transformer at the left of figure 2.16 (possibly an induction coil assembly pressed into service) was used to produce the high voltage needed in such a discharge tube. A typical recorded waveform produced by such apparatus can be seen in figure 2.18 and, in spite of the fuzziness, this is now really beginning to approach the type of display required. The sinusoidal shape of the waveform is readily apparent.

Figure 2.16   Ruhmer's cathode glow oscillograph [29]. Photograph: Science Museum (393/86).

Figure 2.17   Form of gas discharge tube used in Ruhmer's oscillograph [29].

There were some fundamental problems with this method. A discharge tube will not strike unless the voltage between the electrodes exceeds some specific value (the striking voltage) and will be extinguished when it falls below some other value (the maintaining voltage). The result is that dark gaps are left in the trace where the applied voltage is small. These are clearly seen in the trace of figure 2.18. Gehrke pointed out that this defect could

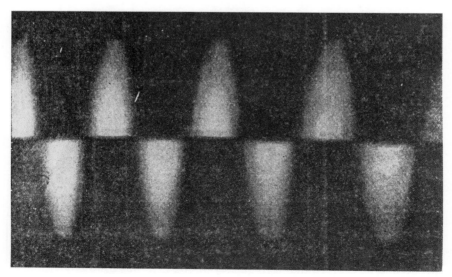

Figure 2.18   The type of trace produced by Ruhmer's oscillograph for a sinusoidal voltage input [29]. Photograph: Science Museum (394/86).

be overcome by superimposing the alternating voltage on a steady DC voltage to avoid extinction of the tube [33]. Another problem was that the glow on the electrodes had thickness as well as length. This also varied with the voltage and introduced some distortion on the photographic record. A later version of the tube used flat-plate electrodes rather than thin round wires, the plates being viewed edge on [34,35]. Such a tube can be seen in the apparatus due to H Boas of Berlin, illustrated in figure 2.19 [36]. The tube had aluminium plates 10 cm long and 15 mm wide and was viewed through the mirror mounted on the shaft of the motor at the bottom left of the picture.

Figure 2.19 Apparatus constructed by Hans Boas, incorporating a flat-plate form of Gehrke tube [36]. Courtesy: Longman Group Ltd.

As an example of the results which could be achieved with apparatus of this sort we may quote some figures published by Herman Diesselhorst, also of Berlin, [37] in 1907. With a concave mirror rotating at 114 revolutions per second he claimed that a frequency of 680 kHz was distinctly visible as an alternating signal, although the curve shapes representing one cycle could not be seen since one period represented only 0.46 mm distance on the photographic plate. When the frequency was 11 kHz one period represented about 27 mm on the plate and the sinusoidal nature of the curve could be seen easily. As a further refinement, Fleming mentioned the use of a modified version of the tube having two anodes and one cathode to permit the observation of two waveforms simultaneously [38]. The present author has been unable to discover any further details of this double tube, but presumably the central cathode plate would have had the two anodes at its opposite ends, and two separate areas of cathode glow would have been created.

Although the glow light oscillographs were in no sense widely used or commercially produced, they do represent a very interesting attempt to extend the well tried principle of spark observation into the realm of what might be termed 'proper oscillography'—i.e. the display of waveforms as real graphs of the variable as a function of time. The comparatively late date of these endeavours (1904–1906) is also noteworthy, since by that time the contact methods and mechanical oscillographs were available. Clearly, there were individualists who, against the apparent odds, were prepared to challenge the accepted methods. We shall encounter more such independent thinkers in Chapter 6.

Before moving away from the spark method, there are one or two other related matters which it might be of interest to mention in passing. An alternative proof of the existence of Leyden jar oscillations was provided by Carl Adolf Paalzow [39–41]. This involved passing the discharge through a Geissler tube. When a undirectional voltage is applied, the greater length of glow appears at the cathode, as pointed out above. When the Leyden jar discharge passed, more or less equal glows appeared on both electrodes thus proving the bidirectional nature of the current. A permanent magnet held near a gas discharge causes it to deflect. By the usual law of magnetic forces, a undirectional current discharge would deflect in one direction only. However Paalzow showed that the Leyden jar discharge glow split up into two distinct parts, one deflected to one side and one to the other, thus making abundantly clear the presence of currents in both directions.

During the first decade of the 20th century, wireless telegraphy made use of the so-called 'musical spark' form of transmission [42]. Spark balls mounted around the periphery of a rotating wheel passed in turn near a stationary electrode and the effect was to produce bursts of damped radio-frequency (RF) oscillation, the burst occurring at regular intervals so that a musical note was heard in the earphones at the receiving station. A rotating-mirror method was sometimes used to determine the frequency of the bursts and also the frequency of the RF oscillation within each burst [43]. A typical arrangement of this sort was the Campbell–Paul photographic spark counter. It consisted of a V-shaped enclosure, the spark occurring at the end of one arm of the V and its image being thrown by a concave mirror down the other arm onto a photographic plate. This plate was moved slowly past the image by a weight-driven mechanism so that a row of sparks would be recorded. The accuracy of the measurement depended critically upon the precision with which the speed of rotation of the shaft could be determined, and a very elaborate system was used for this purpose. A commutator mounted on the shaft was used to switch a capacitor in and out of one arm of a bridge circuit, and by balancing the bridge the speed could be accurately calculated. Frequencies up to about 2 MHz were measurable in this way.

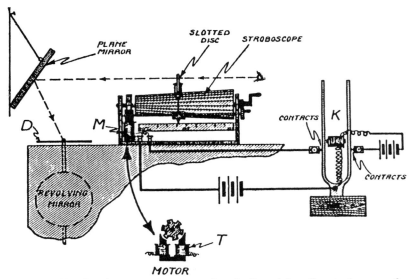

Figure 2.20   Spark counter incorporating the Drysdale roller stroboscope for measurement of the speed of rotation of the mirror shaft [43]. Courtesy: Electrical-Electronic Press.

Another method of determining the speed of a shaft with such apparatus was the instrument known as the Drysdale roller stroboscope, named after C V Drysdale of the Northampton Institute (now City University London) and later Superintendent of the Admiralty Research Laboratories, Teddington [44],—which is shown in figure 2.20. An alternating supply of known frequency, here generated by using a tuning fork to make and break a battery circuit, supplies a synchronous motor which drives a tapered horizontal drum at a speed of 1,000 revolutions per minute. Resting on this drum is a slotted disc which also rotates due to the friction between itself and the drum. By moving the disc horizontally along the drum its speed can be varied, and can be read off the scale mounted on the instrument. A stroboscopic disc (D) mounted on the mirror shaft is viewed through the slots in the disc with the aid of the tilted mirror, and the disc position is varied until the stroboscopic disc appears stationary, and the shaft speed can then be determined. It was claimed that radiofrequencies up to several hundreds of kiloHertz could be measured in this way with an accuracy of 0.1 %.

2.2 Acoustic Measurements

Let us now turn our attention to that other field of endeavour where the need to examine waveforms was acutely felt in the nineteenth century, namely acoustics and phonetics. Mention has already been made of the use by Wheatstone of the rotating-mirror principle to study the phenomenon of the 'singing flame' in a resonating pipe. This idea was extended by R König who, in the 1860s and 1870s, studied the air vibration in organ pipes. Karl Rudolph König was an interesting character; born in Königsberg, Prussia in 1823, he moved to Paris as a young man and learned his trade as a maker of string instruments. He interested himself in acoustics generally, and in 1859 established his own firm as a 'constructor of acoustic instruments'. Many eminent scientists of the day made use of his services. For example, in the Archives of the Institutions of Electrical Engineers, London, there is a copy of a book written by F J Pisko which describes the instruments displayed at the Paris World Exhibition. This particular copy of the book was owned by Silvanus P Thompson and inside it he has written the note 'I bought this book in memory of my old and esteemed friend, Dr Rudolph König, to whom it belonged'.

König used the apparatus of figure 2.21 [45–51]. Gas was supplied to a chamber, one wall of which was a membrane set opposite a hole in the side of the pipe under investigation so that the pressure variations were communicated to the gas in the chamber. The gas then passed to a jet burner, the flame fluctuating in height accordingly. This flame was then viewed through a four-sided mirror drum rotated by a hand crank. This can

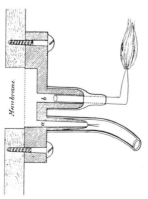

Figure 2.21 König's manometric capsule [45].

be seen in figure 2.22 which shows the complete apparatus arranged for simultaneous study of two pipes. This system, often referred to as the 'manometric capsule', could easily be modified for the study of speech and music sounds by the simple expedient of affixing a speaking tube to the membrane as in figure 2.23. When the flame was viewed in the mirror its image was drawn out to form a trace similar to those shown in figure 2.24. These are sketches, not actual photographs.

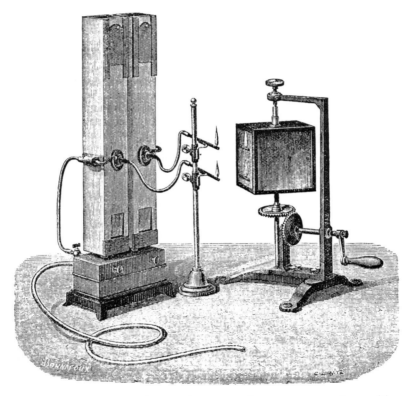

Figure 2.22   Two König capsules mounted on two organ pipes, with viewing mirror [46].

It is tempting to assume that the height of the flame at any instant will be proportional to the pressure in the capsule at that time so that the tip of the flame viewed in the mirror will trace out the required pressure–time waveform. However a closer examination of figure 2.24 will show that this is far from being the case. The 'flame pictures', as they were often called, always have a very curious hooked or sawtooth appearance which is very characteristic. This curving appearance is so marked that in some cases the function represented by the edge of the flame seems to be multivalued,

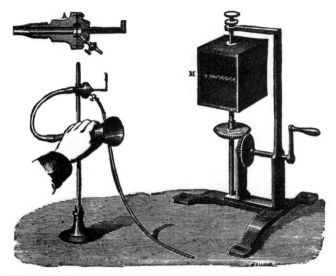

Figure 2.23 Manometric capsule with speaking tube [46].

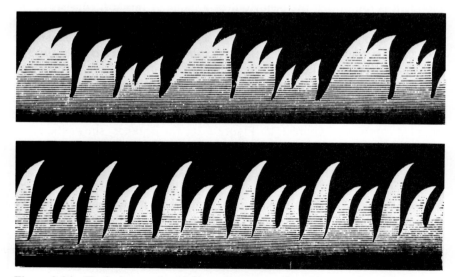

Figure 2.24 Typical examples of flame pictures produced by vowel sounds sung into the apparatus of figure 2.23. (1893 *Ganot's Physics* Translated E Atkinson, 14th edn (London: Longmans-Green) p 273) Courtesy: Longman Group Ltd.

which is clearly an impossibility since only one value of air pressure can exist at any time. Many of the books which describe König's apparatus gloss over these obvious imperfections and concentrate instead on the simpler problem of measuring the pitch of the sound by observing the distance between peaks. Others realised that the phenomena present were very complex. J G Brown, [52] for instance, said in 1911 'It is very evident that the flame-pictures are in no sense time–displacement diagrams of the sound waves producing them. They are overlapping images of successive phases of the vibrating flame and the waveform can only be inferred approximately from such a result. The luminous tip of the flame is not a definite thing which vibrates vertically.'. H Bouasse, in his book *Acoustique Général* (Ondes Aériennes) provided a careful analysis of the mechanism involved [53]. He too pointed out that 'in general there exists no correspondence between the law of variation of pressure and the law of vertical displacement of the extremity of the flame'. He showed that if the diaphragm of the capsule were moving with displacement $x = a\sin(\omega t)$ then the pressure within the capsule would be

$$p = p_0 + k\frac{\mathrm{d}x}{\mathrm{d}t} = p_0 + k\omega a\cos(\omega t)$$

where p_0 is the steady supply pressure of the gas main. In fact if the second term in this expression happened to become larger than the first, negative pressure would result and the flame would be extinguished. When studied closely it is found that the behaviour of a flame of this sort consists of the emission of a luminous puff of gas during compression, followed by a relatively dim period during rarefaction. These luminous puffs rise up in the flame and if they occur with sufficient rapidity it is quite possible for two puffs to be visible at the same time. The puffs also spread outward as they rise which complicates the picture even further. Bousasse was able to confirm the correctness of his deductions by analysing what happens when two sinusoidal pressure variations are applied simultaneously by means of the coupled capsules of figure 2.25. Figure 2.26, for instance, shows the effect of the combined displacement

$$x = \sin(\omega t) + \sin(2\omega t - \varphi)$$

where $\varphi = \frac{1}{2}\pi$. The combined curve exhibits two localised regions of maximum slope, one greater than the other, resulting in flames of unequal height producing the flame picture shown on the right-hand side. He made studies of various combinations of frequencies and phases and in each case he was able to predict correctly the general pattern of peaks in the flame picture. He was also most insistent that the flames were properly referred to as 'pulsating' rather than 'vibrating'. Actually he was most scathing about those who appeared to believe that the flame simply went up and

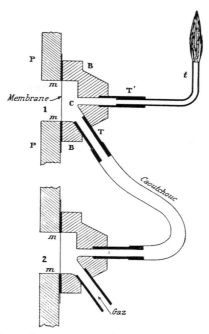

Figure 2.25 Coupled capsules used by Bouasse in his detailed study of flame behaviour [53]. Courtesy: Libraire Delagrave and Dr J Vergnes.

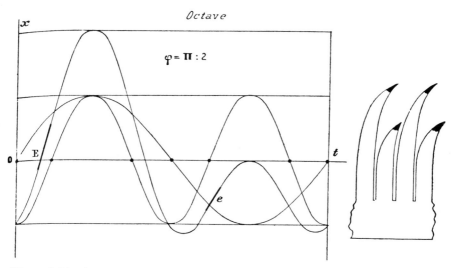

Figure 2.26 An example of Bouasse's analysis, showing the combined waveform $x = \sin(\omega t) + \sin(2\omega t - \frac{1}{2}\pi)$. E and e are the two regions of maximum slope, and the resulting flame picture is sketched on the right [53]. Courtesy: Libraire Delagrave and Dr J Vergnes.

down with the pressure, producing a simple waveform at the tip:

> To show the unutterable stupidity of our pundits, here is the text of Monsieur
> ..., Member of the Institute who, in France, is supposed to know some few
> words about acoustics: 'The flame lengthens and shortens with a movement
> synchronized to that of the internal air'. This stupid remark is printed a few
> centimetres below a Figure which proves exactly the opposite. Those are the
> sort of idiots who rule the destinies of the French university!

Pierre Maxine Henri Bouasse DSc, Professor of the Faculty of Science at
Toulouse, was clearly not a man to suffer fools gladly.

Various attempts were made to improve and adapt the idea of the
manometric flame capsule. One of the main problems in creating permanent
photographic records of the flame was the low actinic value of the light
emitted, coupled with the low sensitivity of the photographic emulsions of
those days. It will be recalled that the first really sensitive gelatine emulsion
plate was offered for sale in 1878 [54]. Earlier dry plates with gelatine or
albumen emulsions were less sensitive, and the older wet-plate processes
involved a great deal more preparation and trouble. The photography of
sparks had suffered from the same insensitivity, and one of the common
remedies there had been to make the spark electrodes out of cadmium, the
evaporation of small amounts of metal into the spark increasing the actinic
value of the light emitted [55]. In the case of the flames it was found that
surrounding the gas jet with oxygen in the manner shown in figure 2.27 produced a much brighter light [56–59], although some experimenters seemed

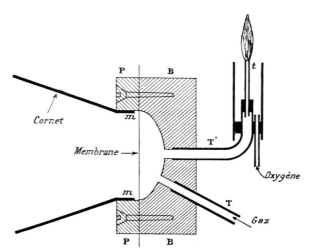

Figure 2.27 An example of a capsule with its flame burning in an
atmosphere of oxygen [53]. Courtesy: Libraire Delagrave and
Dr J Vergnes.

to manage without this [60]. Various mixtures of gases were also tried, coal gas enriched with benzene, or an acetylene–hydrogen mixture being popular combinations.

K Marbe [61,62] in 1906, tried to do away with the need for photography altogether by using a smoky flame issuing from a jet and allowing the stream of smoke to play on a band of paper moving over it, thus leaving a sooty trace upon it. His system is shown in figure 2.28. The membrane of the gas capsule was in connection with the interior of the wooden organ pipe which was set into acoustic resonance by sounding the tuning fork. The pulsating flame produced the upper trace of figure 2.29. The trace below (labelled Figur 4) represents an attempt to use the method for electrical observation and was produced by the town's AC current supply passing through a telephone ear piece which, in turn, actuated the membrane of a König capsule. (Some other arrangements of this sort will be described shortly.) The two traces labelled Figur 5 show simultaneous records of a tuning fork and the AC supply, and those of Figur 6 and Figur 7 show the results of singing vowel sounds into a microphone connected to the telephone ear piece. This very crude technique was refined by J G Brown [63] who started off his researches by making detailed observations of the way in which trails of smoke arose from various parts of the flame. These eventually led him to the use of an acetylene gas flame issuing from a tube

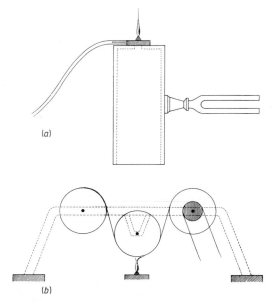

Figure 2.28 Marbe's apparatus for producing smoke traces: (*a*), flame capsule mounted on resonating pipe; (*b*), flame playing on band of paper moving over rollers [61].

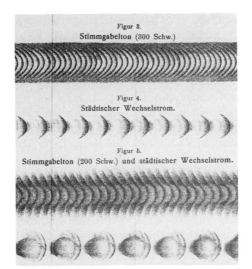

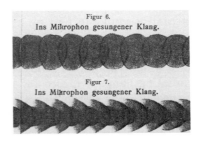

Figure 2.29 Some of Marbe's smoke traces; for details see text [61].

with a 1 mm orifice held at an angle of 65° to the vertical, perpendicular to the direction of travel of the paper band and 2 cm away from it. The first type of trace which he produced is shown in figure 2.30(*a*). As Brown said himself, 'This produced an interesting pattern which characterises the vibration but is difficult of interpretation'. By increasing the gas pressure he was finally able to produce traces such as the one shown in figure 2.30(*b*). This was a trace of the AC lighting mains, again taken using a modified manometric capsule incorporating a telephone diaphragm. Quite a number of people used this type of capsule in an attempt to record electrical waveforms, those of figure 2.31 being typical examples [64–66]. Professor Elihu Thomson in the USA laid claim to having been the first person to use

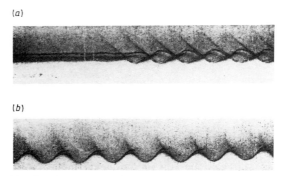

Figure 2.30 Smoke traces of AC mains voltage produced by J G Brown: (*a*), first attempt; (*b*) using increased gas pressure [52].

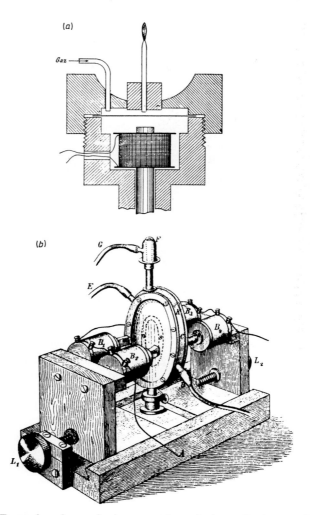

Figure 2.31 Examples of capsules incorporating telephone diaphragms in order to determine the waveforms of electrical signals: (*a*), Bouasse [53], (*b*), Colley [68]. Courtesy: Libraire Delagrave and Dr J Vergnes.

a system of this sort, although it is not completely clear from the account of his remarks quoted in [67] whether the date in question was 1881 or 1887, and so R Colley (in spite of his English sounding name, a Russian who worked in Berlin for a period) may well have done so before him [68]. It is interesting to note also that a capsule of this sort formed the basis of a primitive type of amplifier known as 'Horton's flame amplifier' [69,70]. In this, the flame was made to rise between two inclined metal plates, and the electrical resistance which appeared between the plates varied according to the height of the flame. The ions in the flame provided the conducting medium. A current flowed in a separate battery circuit connected across the plates, and a current gain was obtained relative to the signal current producing the pressure variations in the capsule. Details of the device are scarce, but the waveform distortion must have been pretty awful.

(a)   (b)

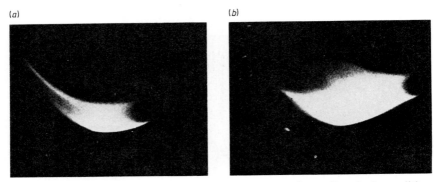

Figure 2.32 Brown's downward-pointing flame: (*a*), in undisturbed condition; (*b*), affected by vibration of telephone [52].

J G Brown also made some further modifications to the gas jet, pointing it downwards at an angle of 45°. The rather shadowy photograph of figure 2.32(*a*) shows the flame burning in an undisturbed condition. When the diaphragm of the telephone was actuated by the alternating current, the flame seemed to spread out as in figure 2.32(*b*). When it was examined carefully with the aid of a stroboscopic disc (a method first introduced by A Töpler of the Baltic Polytechnic in Riga, Latvia, to study the singing flame) [71] he found that the lower edge of the flame was moving up and down in a manner more or less proportional to the pressure variations. He therefore photographed the flame through a vertical slit onto a moving plate and produced the flame pictures of figure 2.33. The first of these shows the waveforms of two sinusoidal currents with two different amplitudes, and the second shows two waveforms rich in third harmonics. Brown's traces represented a distinct improvement on the usual König pictures and seemed to produce outlines which were close to the true waveforms.

Another technique which was suggested was referred to in the literature of the time as the 'transversal' system [72–74] shown in figure 2.34. A funnel R and speaking tube G lead the pressure variations to a small jet situated horizontally near a sensitive flame, and the resulting disturbance to the smooth flow of gas causes the flame to vibrate up and down, these variations being observed in a rotating mirror in the usual way. There does not seem to be any evidence that this particular arrangement was any great improvement over the usual König capsule, and there is only passing mention of it in the literature.

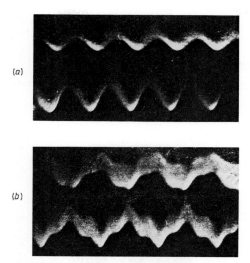

(a)

(b)

Figure 2.33 Traces obtained by Brown with downward-pointing flame for (*a*) sinusoids and (*b*) waveforms containing third harmonics [52].

König also used the manometric flame for purposes of harmonic analysis by combining it with a bank of Helmholtz resonators or resonant chambers as in figure 2.35 [75–79]. His apparatus consisted of eight acoustic resonators each one in connection with its own capsule, the eight flames being mounted in a row, one above the other, and viewed in the usual four-sided mirror. A tone of a given frequency would cause a flame to vary only if the resonator happened to be tuned to that particular frequency. The others would not respond, and when seen through the mirror their flames

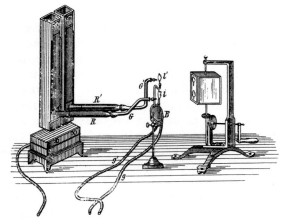

Figure 2.34 The 'transversal' type of sensitive flame [74]. Courtesy: J A Barth, Leipzig.

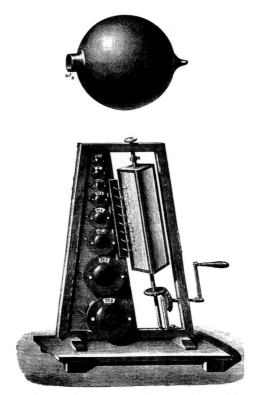

Figure 2.35 A harmonic analyser incorporating eight Helmholtz resonators, each with its own flame capsule (An individual resonator is also shown above.) [46].

would produce only continuous bands of light. The active one alone would produce the serrated line as previously described. The resonators were tuned to the fundamental, second, third ..., harmonics of the note being sounded and any complex sound could be resolved into its harmonics by observing which of the flames responded. With this apparatus, the frequencies of the resonators were fixed and so it was rather an inflexible piece of apparatus, suited to the analysis of tones of one particular frequency. A second version (figure 2.36) had fourteen resonators, each formed out of two telescopic tubes so that it could be tuned over the interval of a musical third. Each one was provided with a scale from which its frequency could be determined. In order to set it up initially to perform the analysis of a particular musical sound a sonometer string was tuned to the fundamental pitch of that sound, and this was used to set the first resonator to maximum response. The sonometer was then used to produce harmonic notes by stopping and bowing in the appropriate manner and the other resonators were tuned in turn. In this way a whole range of fundamental pitches was catered for, and analysis up to the eighth or ninth harmonic was possible in some cases. W Hallock [80,81] constructed a similar system and succeeded in producing a permanent record of the movement of the flames by photographing them through a slit using a camera which was swung around by hand or by a spring mechanism.

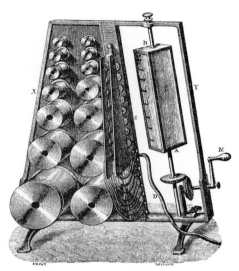

Figure 2.36   The tunable version of König's harmonic analyser [46].

To summarise: the manometric flame was essentially an indirect method of rendering visible the movement of vibrating air masses or diaphragms. The air pressure vibrations were converted into gas pressure changes in the capsule, and these in turn were observed in a rather crude manner by the movements of the flame.

Clearly, much of the uncertainty arose because it *was* an indirect method. The vibration of the membrane was first converted into a pressure variation (according to some approximate and rather ill defined law), and this in turn caused even less well defined variations in the shape of the flame. There would be a much better chance of achieving some sort of fidelity in the inscription if these various intermediate steps could be removed altogether, and if some means could be found of making the movement of the diaphragm, or other vibrating object, draw the waveform directly. Of course, the natural frequency response characteristics of the diaphragm itself would remain, and would still 'colour' the resulting trace, but at least all the other uncertainties in the chain of action would be absent.

Physics textbooks dating from the second half of the nineteenth century usually describe various manifestations of these simple direct-recording devices, and a selection will be given here. The Duhamel 'vibrograph' or 'vibroscope' of figure 2.37 (named after Jean Marie Constant Duhamel, Professor of Analysis at Paris) is a typical example [82,83]. A revolving drum covered in lamp-black or soot was mounted on a threaded axle so that it moved laterally as it rotated. The movement of the tip of the rod, which was set into vibration by means of a violin bow was revealed by the spiral trace inscribed on the drum by the bristle. König constructed a version of this apparatus, shown in figure 2.38, for observing the vibration of the tine of a tuning fork. A second stylus operated by an electrical solenoid could be made to 'peck' at the surface of the drum. This was actuated by a seconds pendulum with a contact which dipped into a pool of mercury once in every swing, thereby completing a battery circuit and providing time calibrations alongside the main trace. The frequency of the fork could be measured in this way if desired.

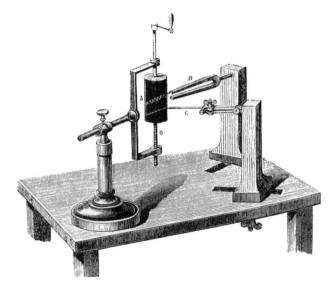

Figure 2.37   The Duhamel 'vibroscope'   [78, p 228].   Courtesy: Longman Group Ltd.

Theodor Stein, a Doctor of Medicine from Frankfurt on Main, in 1876 [84] produced permanent records of the movement of tuning fork tines using the optical method shown in figure 2.39. A tiny hole was bored in the end of the fork and a beam of sunlight, reflected through this hole, was allowed to fall on a photographic plate which was moved along a slide to provide a time axis. He used a similar technique to observe the vibrations of a string—see figure 2.40. A small piece of blackened mica with a hole in it was fixed to the string. A simultaneous record of the movement of several strings could be produced by appropriate mounting of the mica pieces so that one did not obscure the light from the others. Sometimes, in these direct inscription devices, a plate falling down a vertical slide was used to record the waveforms instead of a rotating drum. Since the plate was accelerating rather than moving with constant velocity the extraction of the actual displacement versus time waveform was a somewhat more complicated procedure.

Figure 2.38 König's 'vibrograph' being used to record the motion of a tuning fork. The solenoid arrangement was used to record time calibration marks ([47], p 154). Courtesy: J A Barth, Leipzig.

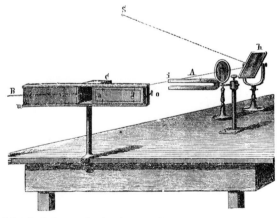

Figure 2.39 Stein's method of recording the vibrations of a tuning fork [84].

Returning to the problem of recording the movement of a diaphragm in response to a sound; one of the best remembered of these direct inscription recording methods was the 'phonautograph' invented by Léon Scott [85–88]. Édouard-Léon Scott de Martinville, to give him his full style and

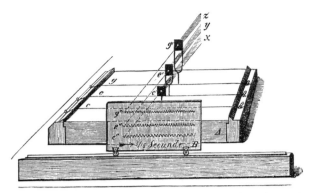

Figure 2.40   Stein's method for recording the vibration of strings [84].

title, was the last remaining descendant of a Scottish family who emigrated to France with the exiled Stuart, James II. A printer's proof corrector by profession, living in Rennes, Brittany, Scott became interested in phonetics and eventually came to the notice of Rudolph König, who assisted him with the construction of his instrument. An engraving of the device can be seen in figure 2.41. It consisted essentially of a funnel, closed at one end with a membrane to which was attached a bristle (o). The bristle touched the surface of the usual smoked drum, and when sounds were spoken into the funnel the vibrations of the membrane left a record on the drum. The phonautograph is widely described in the literature, since it is generally regarded as being the forerunner of Edison's phonograph [89]. Fine engravings usually accompany these accounts, but many of them are vague

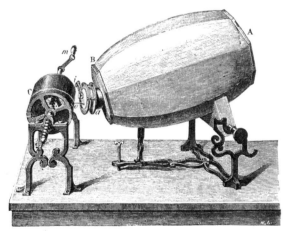

Figure 2.41   Léon-Scott's 'phonautograph' ([78], Section 287). Courtesy: Longman Group Ltd.

as to its exact method of operation—in particular, as to how the in-and-out motion of the diaphragm was converted into a side-to-side record on the drum. The secret lay in the little bracket (i) which can be seen mounted on the funnel near the membrane. A metal point was mounted on this bracket and was pressed against the membrane just above its centre, thereby preventing movement and creating a nodal line of no motion down its vertical diameter. The bristle was mounted on this nodal line just below the centre of the membrane and it was therefore stationary as far as in-and-out movement was concerned. When a sound was spoken into the funnel, the membrane vibrated in the manner illustrated in figure 2.42, the tip of the bristle moving from side to side as a result. This type of motion must have produced a very peculiar sort of waveform and as one writer commented 'its (the bristle's) movements correspond in frequency, though not in any other respect, with those of the air' [90].

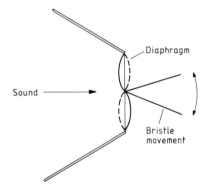

Figure 2.42 Mode of vibration of the diaphragm in the phonautograph (plan view).

The invention of the phonautograph was announced to the world in an article in the French journal *Cosmos* written in 1859 by the French Jesuit, Abbé François Napoleon Marie Moigno who was, in fact, the publisher of that Journal [91]. This article is very short on technical details, but exceedingly long on philosophical ramblings: when discussing speech vibrations in the air, the author goes on at length:

Ah, if I could place upon this air that surrounds me and which conceals all the elements of a sound, a pen, a stylus, this pen, this stylus would create a trace on a suitable fluid layer — but where to find a point of support? To attach a pen to this fleeting, intangible, invisible fluid is a chimera, it is impossible. Wait. This insoluble problem is partly resolved. There exists an inventor, a subtle artist for whom nothing is impossible; it is God. Let us consult Him; let us consider attentively that marvel amongst all marvels, the human ear ....

and much more in similar vein! In an address some years later to the French Academy of Sciences [92] Léon Scott reinforced his claim to having been the inventor of the phonautograph by the somewhat melodramatic procedure of requesting that a sealed packet which had been deposited with the Secretary of the Academy on 26th January 1857 should be opened. This was duly done, and it was found to contain a note entitled *Principes de Phonautographie* describing the instrument. Moigno also exhibited Scott's results at the British Association Meeting at Aberdeen. Prince Albert, who was attending the meeting, was very taken with this work and insisted on taking some of the curves to show to Queen Victoria who was also in Scotland at the time.

In 1877, Alexander Graham Bell, of telephone fame, read a paper to the Society of Telegraph Engineers in London [93–96]—a lecture which was repeated in Sydney [97], and doubtless in many other places as well since a man of his eminence would have received numerous invitations to speak at learned societies around the world. In the course of this talk he described the events which led him to construct an improved version of the phonautograph. One of the students at the Institute of Technology in Boston, a Mr Maurey, had managed to increase the amplitude of the recorded trace by causing the diaphragm to operate the recording stylus through a light wooden lever, one foot in length. When Bell saw this, he was struck by the likeness between it and the way in which the bones of the inner ear were moved by the vibration of the ear drum. He therefore decided to construct his own model of the phonautograph, modelled as closely as possible on the ear and to this end he enlisted the aid of Dr Clarence J Blake of Boston, an eminent ear specialist, who suggested that Bell should go the whole hog and actually use bones taken from a human ear. Dr Blake prepared a specimen for him, lubricating it with a mixture of glycerine and water. With a piece of hay fixed to it to act as a stylus he was able to draw the waveforms of sung notes etc. on a piece of smoked glass moved along beneath it. Two versions of engravings of this apparatus are to be found in the literature—one is reproduced in figure 2.43. This shows the instrument in a general sort of way but it is not really possible to work out in detail the disposition of the bones and how they were connected between the diaphragm and the stylus. The printed version of the lecture gives no further details, but according to Bell himself it was a study of this device which eventually led him to the invention of the telephone. Dr Blake wrote several articles about the instrument in the American journal *Archives of Opthalmology and Otology* in 1876–1879 and, as well as describing his own experiments, he makes reference to work of a similar nature which had been carried out by several other people [98,99]. Léon Scott himself, in the address to the French Academy already referred to, mentioned that he had heard of the use 'by a foreigner of the middle ear of a decapitated dog',

Figure 2.43 The Bell–Blake improved phonautograph using bones from the human ear [94,95].

although he gave no exact reference. The improved model he himself was describing on that occasion used a chain of artificial ear bones.

Another early attempt to register the movement of a diaphragm was the 'logograph', invented by W H Barlow, a distinguished Civil Engineer who designed, among other things, the train shed at St Pancras Station, London, and the reconstructed Tay railway bridge built to replace the ill fated structure of Sir Thomas Bouch. The 'logograph' was described in the Popular Science Review of 1874 and also in the proceedings of the Royal Society [100,101]. No diagram of the machine is shown, but it seems to have consisted of the usual trumpet and membrane arrangement, a spring being attached to the membrane to keep it in a state of tension and also being connected to a pointer with a marker at the end. The marker in this case was a fine sable brush with a glass tube full of ink surrounding it, being so arranged that the ink could seep out into the brush and hence make a mark on a band or drum of paper. Dr Clarence Blake's papers are entitled 'The use of the membrana tympani as a phonautograph and logograph', which leads one to suppose that there must have been some subtle difference between the two instruments, since it states 'phonautograph *and* logograph', rather than 'phonautograph *or* logograph'. He does not elaborate further however and the only clue is his statement referring to 'a logographic curve made with the plate moving at the rate of five millimetres in the second, and finally a phonauto–logographic tracing made with the plate moving at the rate of five centimetres in the second'. Perhaps the logograph gave an overall 'envelope' picture of the sound whereas the phonautograph gave a proper waveform—although this is only conjecture on the part of the present author.

Another Blake, this time Professor Eli Whitney Blake, sometime Professor of Chemistry at Vermont and Professor of Physics at Cornell and Brown Universities, published details in 1878 of 'A method of recording articulate vibrations by means of photography' [102,103]. In his method, a 2.75 in diameter iron disc of the type used for telephone receivers was attached by a stiff steel wire to a hinged mirror. A heliostat was used to direct a beam of sunlight onto the mirror, and the reflected beam was brought to a focus, through a lens, onto a horizontal sensitive plate. This plate was carried on a four-wheel trolley, and was moved along by weights on a string passing over a pulley. A permanent weight was provided to overcome friction; a second weight was added to accelerate the trolley, but was removed just before exposure so that a constant velocity was produced during the recording time. Some very nice curves resulted when sounds were spoken or sung into the diaphragm, but it was obviously a very insensitive device, and Blake himself realised the frequency limitations inherent in such an arrangement. In his own words—'Are all the elements of speech traceable in these records? In other words, is the record complete?'. He tried to resolve the question by connecting the mirror to an ordinary telephone receiver: 'The resulting record was almost a smooth line, showing but very slight indications of the movement of the mirror. It would therefore appear that there are distinctly audible elements which are too minute to be recorded by this method'.

A few years later, in 1891, Oscar Frölich, a Swiss-born engineer working at the Siemens and Halske laboratory in Berlin, attempted to use a somewhat similar method to register the movement of an ordinary telephone diaphragm [104–108], but he mounted a piece of mirror on the diaphragm itself as in figure 2.44(*a*). It will be noted that the mirror is not mounted at the centre of the diaphragm, but is offset towards the edge so that the normal flexure of the diaphragm results in an angular rotation of the mirror, and it is this movement which is recorded on the screen S by the beam from the arc lamp. Frölich attempted to determine the influence of the response of the telephone by a Lissajous' figure technique, using an organ pipe. The spot of light was deflected in one direction by a piece of mirror mounted on a membrane vibrated directly by the sound from the pipe, and in the orthogonal direction by the mirror on the telephone diaphragm, the sound having been picked up by a microphone. His conclusion was that '... the electric oscillations are greatly modified by the telephone, and are very complicated; every simple current put into the telephone produces a sound composed of many different tones'. He also studied the waveform of the mains supply by reflecting the beam of light onto a twelve-sided mirror drum, this serving to produce a time-sweep across the screen. The speed of the mirror was synchronised with the mains so that a steady trace was produced. The natural resonance of his diaphragm is well illustrated by the trace of figure 2.44(*b*), the electrical waveform in this case being a battery current switched on and off abruptly at regular intervals. The curves in Frölich's papers are not calibrated, but one is told that each current pulse lasted for $\frac{1}{10}$th of a second so it is possible to infer that the resonant frequency was about 80 Hz. He states that, knowing the natural frequency and the damping of the diaphragm, it is possible to determine the *actual* shape of the current curve, but he gives no details of this procedure.

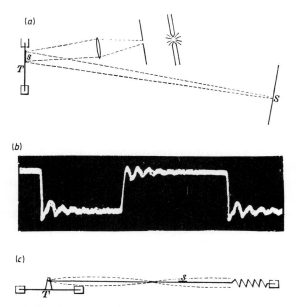

Figure 2.44 Frölich's diaphragm apparatus: (*a*), original version; (*b*), response to a 0.1 s on/off pulse; (*c*), Melde's modification [105,106].

The sensitivity could be improved by a method suggested by someone called Melde (probably Franz Emil Melde, Professor of Physics and Astronomy at Marburg—no further references are given in Frölich's papers) and illustrated in figure 2.44(*c*). A string under tension is fixed to the centre of the diaphragm in the manner shown, and vibrates with a node at its centre. The angular displacement of the piece of mirror, s, will then be comparatively large. This idea is not analysed in detail, but the string would certainly have superimposed one or more enormous resonant peaks upon the frequency response of the system. Various curves are reproduced in Frölich's paper published in the *Electrician*, but it is impossible now to assess how accurately these followed the actual current waveform—not very well one is led to believe by various comments made in the literature [109–111]. One sentence in his own paper might be regarded as having a rather ominous ring: '... the determination in this manner of the phase differences in alternating currents whenever an iron core exists gives results quite different from those to be expected from theoretical considerations'. Professor Blondel, [112] whose work will be referred to extensively later in this book, went so far as to point out that the telephone was really the last instrument which should be employed in the determination of current waveforms since it: (*a*) has a very complicated mode of vibration, (*b*) has insufficient damping and (*c*) has self-inductance and hysteresis which completely deform the curves. The significance of these factors will become apparent in a later chapter when Blondel's analysis and conclusions are considered in detail.

A similar sort of device intended for electrical waveform observation was constructed by the distinguished American engineer Elihu Thomson in 1888 [113]. His English parents moved from Manchester to Philadelphia with their ten children when Elihu was five years old. He eventually became a partner of E J Houston, forming the Thomson–Houston Company, a predecessor of the General Electric Company. He was the recipient of over 700 patents and is credited with the discovery that by breathing a mixture of helium and oxygen, underwater divers could avoid the 'bends'. Thomson started his search for a waveform recorder by building the instrument shown in figure 2.45 in order to record alternating current waveforms. The electromagnet of figure 2.45(*a*) was constructed so as to have an annular gap between its poles. In this gap, a light coil was suspended on the end of a lever as in figure 2.45(*b*). The far end of the lever carried a pencil point resting on a strip of paper wrapped around the usual type of screw-axled cylinder. Although he managed to produce some current and voltage curves with this, its drawbacks must have been very obvious because he abandoned the idea and went on instead to construct the diaphragm apparatus shown in figure 2.46 [114,115]. Two coils (c c) were wound on a bundle of iron wires, one carrying DC to polarise the core, the other carrying the AC under investigation. From the middle of an iron diaphragm situated near the core a short aluminium wire, flattened out at one point to produce a flexible hinge transferred the motion to the lever L with its fulcrum at F. The magnified movement was transferred to the pivoted mirror M through the wire loop G, the mirror being tensioned to rest against it at H by means of the spring s. The use of aluminium wire with a flattened portion in this way to provide a hinged joint without any 'lost motion' due to backlash is perhaps the most notable feature of the design. A beam of light was caused to fall upon the mirror, and thence onto a screen situated at a distance of twenty inches. The whole thing (apart from the screen) was arranged to

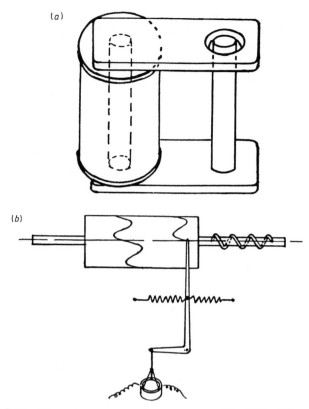

Figure 2.45 Current wave recorder constructed by Elihu Thomson: (*a*), magnet assembly with annular gap; (*b*), moving-coil and recording arm [67].

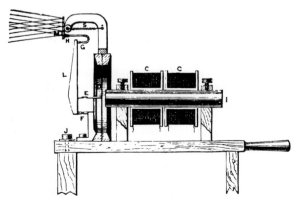

Figure 2.46 Elihu Thomson's diaphragm waveform recorder [114].

swing horizontally about the vertical pivot J and, by seizing the handle and swinging it around, a waveform could be made to appear on the screen. A double version of this instrument was also constructed with the mirror mounted on gimbals. One diaphragm was arranged so as to cause up-and-down motion; another moved it from side to side. In this way Lissajous figures could be created on the screen and the phase differences between two alternating waveforms could be determined. Once again, the response of the moving system was neither well defined nor easily calculated and little credence could be placed upon the fidelity of the waveforms produced. The coils presented a large inductive reactance to the circuit and, to quote one author of the time, 'it is difficult to prevent perceptible retardation of the current around the soft iron core' [116]. It must also be mentioned that all telephone diaphragms suffer from the disadvantage that the force on the diaphragm is proportional to the *square* of the current flowing in the coils.

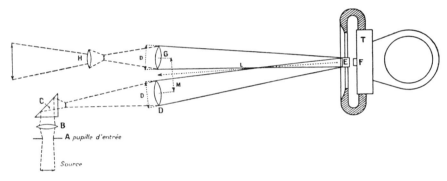

Figure 2.47   Guyau's attempt to use an ordinary telephone ear piece, measuring its movements optically by means of interference fringes [117].

Monsieur M A Guyau in 1913 attempted to register the movement of the diaphragm of an ordinary telephone receiver—see figure 2.47 [117]. He made use of interference fringes created in the thin layer of air between a fixed silvered mirror F and a semi-silvered mirror E on the diaphragm. These fringes were recorded by projecting them onto a film through a horizontal slit, the film being wrapped around a rotating cylinder. After Guyau's original paper, nothing more is heard of this apparatus which the inventor had christened 'l'oscillographe interferentiel', so one must assume that it was not a great success.

Diaphragm waveform recorders for speech and music sounds were brought to a high degree of perfection in the USA by Dayton Clarence Miller who, in 1909, introduced his 'phonodeik', the name meaning 'to show or exhibit sound' [118,119]. The method of operation of the phonodeik can be seen in figure 2.48. A thin glass membrane (D) is situated at the end of a tin speaking trumpet, and to it is attached a fine silk thread or platinum wire kept in tension by a spring (S). The thread is wound several times around a steel spindle mounted in jewelled bearings; the spindle and bearings are shown in figure 2.49. As the membrane moves in and out the mirror is deflected by an amount proportional to its displacement, and this movement is recorded on a moving film in the usual way. The complete instrument can be seen in figure 2.50. Some details of the dimensions of the moving parts might be of interest here. The pulley was 0.6 mm in diameter; the mirror area was 3 mm^2 and its mass was a mere 0.7 mg. No oil was used in the bearings as this would have necessitated space being left between the moving surfaces, and this would have caused undesirable 'slop' and backlash. A displacement of 1/1,000th of an inch in the diaphragm caused an angular movement of 4.85° and an overall magnification of movement of 3,000 times was produced. A displacement of 1/100,000th of an inch was said to be visible in the final trace. In slightly modified form the instrument was also able to project waveforms of sound onto a screen for public viewing, and a portable version having a revolving mirror and a stationary film (figure 2.51) was also available. This was clearly a well engineered instrument of considerable refinement, but even so, the resulting traces were still subject to the usual distortions imposed by the dynamics of the mechanical moving parts and also due to play in the bearings.

This latter factor was minimised, it was claimed, in the version shown in figure 2.52 due to J E Shrader [120]. By using four wires and three springs 'friction at the pivots is avoided and there is no lost motion at their bearings'. Some typical phonodeik traces can be seen in figure 2.53. It is impossible now to judge the accuracy of these traces, but it must be said that they look reasonably convincing and there appears to be a great deal of fine detail present. Professor Miller was a most eminent scientist—President

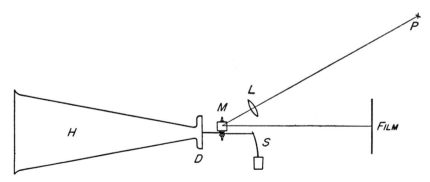

Figure 2.48   Illustrating the principle of operation of Miller's phonodeik [119]. Courtesy: Macmillan & Co.

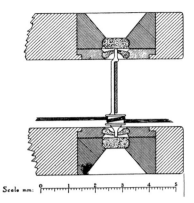

Figure 2.49   The phonodeik; details of mirror shaft and bearings [119, p 19]. Courtesy: Macmillan & Co.

Figure 2.50   Miller's phonodeik [85, p 80]. Courtesy: Macmillan & Co. Photograph: Science Museum (400/86).

Figure 2.51 Portable version of the phonodeik [119]. Courtesy: Macmillan & Co. Photograph: Science Museum (399/86).

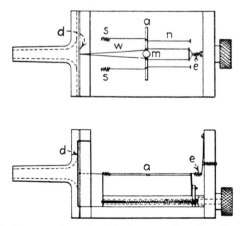

Figure 2.52 Improved antibacklash mounting for phonodeik mirror due to Shrader [120]. Courtesy: McGraw-Hill Book Co.

of both the American Physical Society and the Acoustical Society of America—and was a man of wide interests, being an accomplished flautist and an expert on architectural acoustics. He also became well known for his work on the Michaelson–Morley aether drift experiments. He applied his phonodeik to the problem of the location of enemy artillery during the 1914–1918 war.

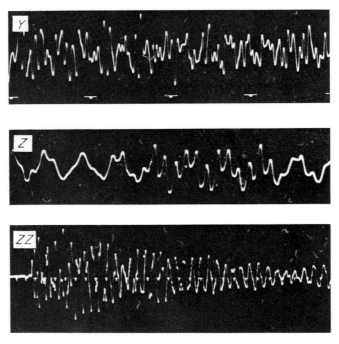

Figure 2.53 Some typical phonodeik traces: Y, brass band music; Z, the clang of a bell; ZZ, a sky rocket [119]. Courtesy: Macmillan & Co.

Another version of a simple diaphragm recorder shown in figure 2.54, was constructed at the research laboratories of Adam Hilger Ltd, by Twyman and Dowell in 1922 [121,122]. This was the so-called 'optical sonometer'†.

There was no new principle involved here, but the most noteworthy feature was the diaphragm which was stretched across the speaking funnel

†This was a slightly unorthodox use of the word, since most scientists will remember, with varying degrees of frustration, the sonometer of the school physics laboratory, which was a tuned string instrument used to determine the wavelengths of musical sounds, and to demonstrate the presence of harmonics, nodes and antinodes by observing the behaviour of paper riders placed on the string.

in the usual way. This was constructed out of a very thin film of celluloid, said to be less than one wavelength of light in thickness, and weighing only 0.2 mg. This was given a thin coating of gold or silver to render it reflecting and a light beam, offset from the centre, was allowed to fall on it, the movement of the reflected beam being recorded on photographic film affixed to the edge of a rotating drum. One to six seconds of speech could be recorded in this way.

Figure 2.54 The 'optical sonometer [121].

Perhaps the ultimate sophistication in diaphragm instruments is represented by the apparatus constructed by Rousselot to enable him to make detailed phonetic studies of the various dialects of the French language. The date of this is uncertain, but it is described in a letter written by J D Boeke in December 1894 and published early in 1895 [123,124]. It was really a five-channel recorder, each channel being provided with a transmitting membrane, a tube, and a receiving membrane with a recording stylus. A sixth trace was made on the smoked drum by a tuning fork for purposes of calibration. The apparatus can be seen in figure 2.55(*a*). One speaking tube was provided with a membrane on a drum strapped near the Adam's apple in order to record the fundamental excitation of the sound. Another drum was strapped under the chin to record (approximately) the movement of the tongue. Two others were connected by levers to the upper and lower lips to record mouth movements, whilst the final tube was supplied with tiny membranes which were pushed up into the nostrils to investigate the effects of nasalisation of the various speech sounds. The final multiple record left on the drum was inspected so as to furnish clues about the pronunciations used in various dialects — with what success is not stated. Boeke expressed his doubts that 'anyone would be able to decipher these hieroglyphics drawn upon the cylinder'.

(a)

(b)

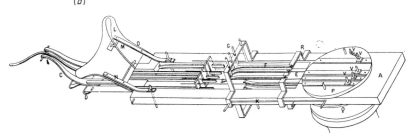

Figure 2.55 Two complicated devices for the analysis of human speech sounds: (*a*), Rousselot [123]; (*b*), Guerot [125].

As a final footnote to this section which has been concerned with the direct inscription of speech sounds, we may examine a device which was described in the French journal La Lumière Électrique in 1881 by Auguste Guerot, who was in fact the editor of the journal. This was a speech recorder attributed to someone of the name of Gentilli from Leipzig [125]. It was somewhat similar in purpose to Rousselot's apparatus, consisting of a series of levers held in the mouth to measure lip positions, tongue configurations, etc. The actual device is shown in figure 2.55(*b*). The curved lever at the left registered the tongue movement. Levers M and N, protected by the shaped moustache guard, registered the movements of the lips. L was a very light flap which registered the passage of air flowing down the nose (nasalisation). The movements of these levers were communicated to the electrical contact strips mounted on the main block (A). There seemed to be a proposal that the resulting signals could be transmitted over wires so that the sound could be recreated at some distant place. It can thus be considered as a proposal for an early mechanical analysis–synthesis Vocoder speech transmission system! Guerot concluded, however, that no one was going to be able to speak at all with that contraption stuck in his mouth!

2.3   The Phonograph

The arrival of Edison's phonograph in 1877 was an event which caused considerable excitement amongst those who were interested in phonetic measurements. It had been appreciated that a membrane or diaphragm, loaded with a pointer and subject to the friction of the pointer on the recording drum, was bound to superimpose its own frequency response characteristics upon those of the sound being analysed. There was no way of knowing whether the recorded trace was a faithful representation of the sound or whether it had been 'coloured' beyond all recognition by the dynamics of the recording process. Edison's phonograph added a new dimension of certainty to the measurements because it not only made a record of the vibrations of the diaphragm but it could also reproduce the sound. The fidelity of the recording could then be verified by the simple expedient of playing it back and listening to it. Fleeming Jenkin and Ewing, for example, whose work will be described later included in their paper the following sentence:

> As a sort of gauge of the merit which our phonograph possessed as a means of recording and reproducing sound, we may mention that no difficulty was felt by hearers in making out sentences as repeated by it, which had originally been spoken in their absence.

Whilst this by no means implies high-fidelity recording, it was at least some reasonable proof that the impression on the record was a fair representation of the original sound.

The recording process was, of course, subject to all the distortions previously mentioned (resonances, stylus friction, etc), but the waveform information could be extracted by playing back the record very slowly so that at least any added distortion caused by a reproducing diaphragm could be avoided. Various experimenters devised quite elaborate mechanisms for the measurement of the tiny indentations on the cylindrical phonograph records, and later on flat discs. For those who may not be familiar with the early history of sound recording, it should perhaps be explained that the cylinder phonograph employed hill-and-dale recording where the recording stylus made indentations of varying depth in the recording medium which, in the earliest models, was a strip of tin foil wrapped around a cylindrical drum. Later models used wax cylinders to record the impressions. Flat-disc recorders—'gramophones' in British parlance—used lateral or side-to-side modulation in grooves of constant depth (apart from some hill-and-dale discs produced by the Edison Company).

In a paper to Nature in 1878, [126,127] Alfred M Mayer, Professor of Physics and Director of the Physical Laboratories at the Stevens Institute, New Jersey, described the general construction of the newly invented phonograph and gave an account of his own attempts to trace the waveform of the recorded sound using a small lever, one end bearing the reproducing stylus, the other having a fine copper point which drew out the profile of the groove on a moving smoked plate. Whilst making the trace, the phonograph cylinder was rotated slowly and uniformly beneath the stylus. He published the diagram shown in figure 2.56. The upper line shows the pattern of indentations observed by eye on the tin foil for the vowel sound A as in B*A*T. The middle curve is the trace produced on the smoked plate and, for purposes of comparison, the lower curve shows the appearance of the flame from a König capsule. Although all three traces are very crude, close examination will show that there is at least some degree of correlation between them. Mayer also suggested another method of obtaining the waveform which was to 'back the tinfoil with an easily fusible substance (wax one supposes) and then cutting though the middle of the furrows we obtain a section in which the edge of the foil presents to us the form of the elevation and depressions'. In other words, he proposed to cut along the length of the groove in order to examine its vertical profile. There is no evidence to suggest that he actually succeeded in carrying out this delicate operation with any success.

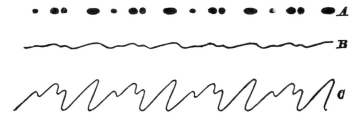

Figure 2.56 Mayer's attempt to trace a waveform from the surface of a tin foil phonograph cylinder [126].

In the same year, Fleming Jenkin and Ewing used a much more sophisticated lever system to enlarge the indentations in the tin foil by a factor of 400 times [128,129]. Working together here were two of the most prominent scientific men of the Victorian era. Sir James Alfred Ewing, a native of Dundee, was at various times in his career Professor of Mechanics at Cambridge, Director of Naval Education for the British Admiralty and from 1916, Principal of the University of Edinburgh. He is, perhaps, chiefly remembered for his studies of ferromagnetic materials and his introduction of the word 'hysteresis'. Henry Charles Fleeming Jenkin (it is pronounced 'Fleming' for some impenetrable reason) was born at Dungeness but received his early schooling at Edinburgh. He was a close friend of William Thomson (Lord Kelvin) and collaborated with Maxwell, Carey-Foster, Latimer-Clark and Wheatstone. He was greatly involved in the early experiments on undersea-cable laying and also with the British Association Committee on Electrical Standards. As a final distinction, he could claim the author Robert Louis Stevenson as a friend and former pupil.

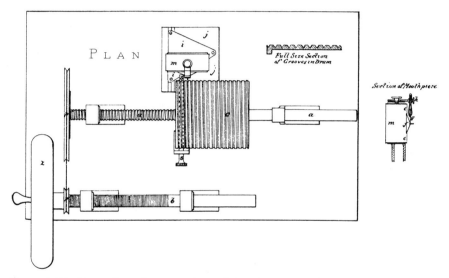

Figure 2.57 Recording phonograph used by Fleeming Jenkin and Ewing [129]. Courtesy: Royal Society of Edinburgh.

The recording apparatus used by these two notable colleagues may be seen in figure 2.57 and their measuring instrument in figure 2.58. The upper part of figure 2.57 shows a normal tin foil recorder consisting of a drum (c) having a spiral groove cut in it mounted on a screwed axle, the pitch of the axle screw being the same as that of the groove so that when rotated, the groove would pass smoothly beneath the fixed recording stylus. A band of tin foil was wrapped around the drum and secured at its edges with shellac varnish. The recording diaphragm itself is shown in greater detail in the small inset drawing (right) and consisted of a cone of stiff paper. The stylus fixed to its centre was made of hard steel, shaped like a chisel, slightly rounded, but sharp in the horizontal plane. In the usual tin foil machines a flywheel was mounted on the shaft in order to keep the speed

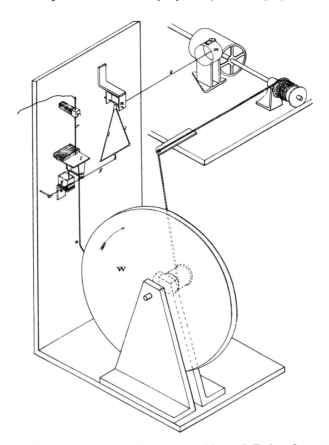

Figure 2.58   Apparatus of Fleeming Jenkin and Ewing for making measurements from the surface of a phonograph cylinder [129]. Courtesy: Royal Society of Edinburgh.

fairly constant and to assist the rotation against the friction of the stylus. The energy stored in a flywheel depends on the square of its angular speed and to render it more effective Fleeming Jenkin and Ewing used the modified arrangement shown. The flywheel was mounted on a separate shaft connected to the main shaft through a belt and pulleys, the diameters of the pulleys being such that the flywheel rotated four times as fast as the recorder shaft. Since lateral motion was also present, it was necessary to provide the second shaft with a quarter-pitch screw thread so that the two pulleys were kept in alignment. The speed of rotation of the recording drum was approximately three revolutions per second during recording.

In the measuring apparatus of figure 2.58 the reproducing diaphragm and stylus were connected by a glass fibre thread (k) to a triangular lever system made out of three light straws pivoted at n. This in turn was connected by a silk thread (y) to a fine glass siphon tube (o), the upper end of which was dipping into a trough of ink (q). The lower end of the siphon was near the periphery of the wheel (W) around which was wrapped a thin band of paper. The thread was held taut by a rubber band (represented here as a spring) fixed to a nail on the mounting frame. The tension in the system was adjusted so as to make it as nearly as possible 'dead beat' in its movement and free from natural oscillations. The wheel was linked to the phonograph shaft by a cord and pulley so arranged that when the recorder shaft was turned slowly, the wheel also turned. An electrical potential difference generated by a frictional machine was maintained between the wheel and the ink via the contact r, the result being that the trace was recorded as a fine series of spots of ink attracted onto the surface of the paper. This arrangement resulted in the minimum friction in the recording system. As previously stated, a magnification of the indentations on the record of 400 times was achieved, and the time axis was magnified about seven times—i.e. the length of the trace on the paper band was seven times the equivalent distance along the phonograph groove. The authors analysed some traces to produce Fourier series for various vowel sounds, and their results were presented in a series of papers to *Nature* in 1878 [130]. A correspondent to that journal was complimentary to the traces, saying [131]

> They differ considerably from the phonautographic speech curves of Léon Scott and König which only succeeded with the vowels, and from the logographic speech curves of Mr Barlow which only succeeded with the consonants, in so much as they succeeded with both.

The tin foil phonograph was a very crude and unsatisfactory device in many respects, not the least being the unavoidable impulse every time the reproducing stylus crossed the join in the tin foil. It was also very difficult to reposition the foil correctly once it had been removed since the rows of indentations had to be realigned accurately with the groove in the surface of the drum. These defects were overcome with the introduction of the removable wax cylinder.

In 1891, J D Boeke [132] made a very painstaking microscopic examination of the grooves, his arrangement being shown in figure 2.59. Some of the drawings he made representing the vowel sounds are also illustrated. His general method of working was to measure the width of the indentations on the record (H) at regular intervals, aided by the pointer (W) and the angular setting scale (P). Since the shape of his cutting stylus was known, these measurements could be translated into depth of cut, and hence into movement of the diaphragm. This sounds very reasonable until one considers the difficulty of locating precisely the edges of the groove. A slightly different approach to the problem was made by J G McKendrick and others [133] who coated the wax record with a solution of collodion

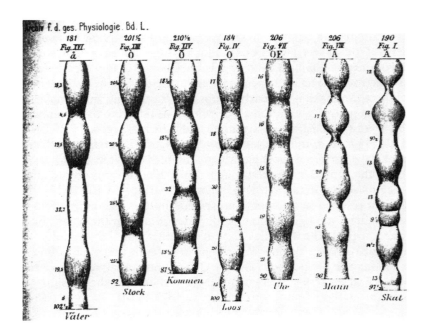

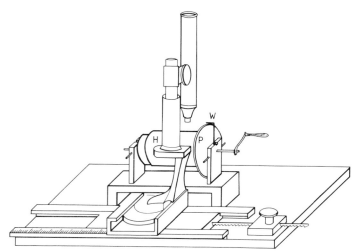

Figure 2.59 Boeke's microscope [132] for the measurement of phonograph indentations. Examples of typical grooves are sketched above.

dissolved in ether. When this had hardened it could be peeled off and laid out flat for examination under a microscope. However, the unrolling process obviously introduced distortions and, as far as can be determined from reports in the literature, it does not seem to have been a great success. Boeke tried to take impressions from the cylinder using very fine tin foil 'such as is used to cover chocolate'—rather like the children's pastime of rubbing an image of a coin onto a sweet paper. This particular piece of information was supplied in a letter from Boeke to McKendrick, but no indication is given as to how successful, or otherwise, this technique proved to be. One suspects that sufficient rubbing pressure to make a good impression would have distorted the original wax recording. Nothing further is heard of the idea, so we can draw our own conclusions. McKendrick made many detailed studies of cylinders, using photographs of the grooves taken through a microscope as a measurement aid, but his general conclusion about the technique was that 'it is not of great value as it does not give the forms of the curves represented by the bottoms of the depressions made by each vibration of the disc (diaphragm) of the phonograph'.

In 1895, C J Rollefson (who is also variously referred to in the literature as Rollesson or Rolleson for some reason) used the wax phonograph to study alternating current waveforms [134,135]. The diagram of figure 2.60(*a*) shows his recording apparatus, consisting of a glass diaphragm (b) to which was fixed a small permanent magnet (a). The movement of the diaphragm was transferred to the recording stylus (f) through a small lever. The current to be studied was applied to the coil wound around the core of iron wires (c). Alternate attraction and repulsion of the small magnet caused

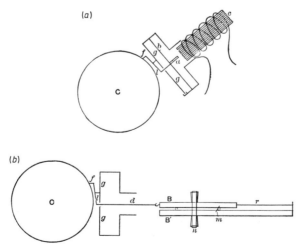

Figure 2.60 Rollefson's apparatus for recording and measuring electrical waveforms by recording on a phonograph cylinder [123].

the recording to be incised on the rotating wax cylinder (C). No figure is given for the recording speed, so it must be assumed that it was the usual commercial speed of about two revolutions per second. The wax record was then transferred to the measuring instrument of figure 2.60(b). B and B' were two small brass bars, the lower one being firmly fixed to the casing of the machine (g). The upper bar rested on two small rollers made of fine iron wire, the whole assembly being held together by the rubber band (n) passing around both bars. The upper bar was connected to the stylus (f) through the lever and the fine wire (d), another rubber band (r) keeping the whole thing in a state of tension. Movement of the diaphragm caused the upper bar to slide to and fro and also caused the rollers to rotate. This rotation was measured by means of a light beam and a small mirror (m) fixed to one of the rollers. A mechanical arrangement was also provided to rotate the cylinder successively through very small angles, but this is not shown on Rollefson's diagram. No indication is given in the author's account of the degree of magnification provided by the system, but he does say that 'the movement of the glass diaphragm when the curve is recorded on the wax cylinder may be made so small that the indentations produced on the cylinder are scarcely visible to the naked eye'. No specimen curves were published, but he reported that 'the results obtained seemed to be very promising'. There is a further comment to the effect that 'By the aid of König's apparatus the number of harmonics present may be determined, and perhaps their relative intensities. Thus by plotting the curve, their relative positions may be determined'. In other words he envisaged that König's harmonic analyser (figures 2.35 and 2.36) was to be used to establish the magnitudes of the harmonic components, the phonograph record then being examined to establish their relative phases—rather a difficult process one imagines for a waveform of any complexity, but perhaps feasible for a waveform having just one harmonic (a fundamental and third harmonic maybe).

In 1889, Ludimar Hermann, Professor of Physiology at Königsberg, [136] attempted to measure the displacement of the reproducing diaphragm of a phonograph by simply fixing a small mirror to it and observing its angular movement with a light beam in the usual way. This idea was taken further in the apparatus shown in figure 2.61(a) which was constructed by Louis Bevier in 1900 [137]. The diagram is fairly self-explanatory, the magnification of the groove profile being achieved by a combination of the lever connecting the stylus D to the tilting mirror F and the use of a long light beam. During the measuring process the cylinder's speed was reduced to one revolution in nine minutes. One second's worth of speech produced a trace 35 ft long and the groove profile was magnified 1,000 times. The author was of the opinion that 5,000 times would have been possible without too much difficulty and indeed in a later paper [138] entitled 'The vowel A^e (as in H*A*T)' he described how the modified measuring head of figure 2.61(*b*) using multiple levers and a light beam 186 inches long was able to provide a magnification of 5,952 times. In subsequent papers he published a series of studies of the various different vowel sounds [139].

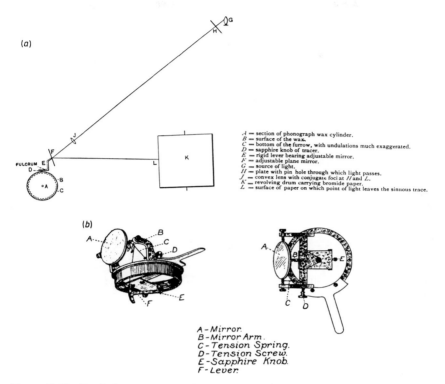

A = section of phonograph wax cylinder.
B = surface of the wax.
C = bottom of the furrow, with undulations much exaggerated.
D = sapphire knob of tracer.
E = rigid lever bearing adjustable mirror.
F = adjustable plane mirror.
H = plate with pin hole through which light passes.
G = source of light.
J = convex lens with conjugate foci at *H* and *L*.
K = revolving drum carrying bromide paper.
L = surface of paper on which point of light leaves the sinuous trace.

A - Mirror.
B - Mirror Arm.
C - Tension Spring.
D - Tension Screw.
E - Sapphire Knob.
F - Lever.

Figure 2.61 Bevier's apparatus for making measurements of phonograph grooves:
(*a*), principle of operation of simple system;
(*b*), multiple-lever head developed later [137, 138].

In his Chairman's address to the Royal Society of Edinburgh for the season 1896–1897, Professor McKendrick, Professor of Physiology at Glasgow, described his improved method of measuring the indentations on a phonograph cylinder of the normal commercial type [140]. The apparatus is shown in the drawing of figure 2.62. This is labelled as having been drawn by Dr William Snodgrass who must have been, one assumes, one of McKendrick's colleagues. The phonograph itself was basically a normal commercial machine, but the usual clockwork drive was replaced by the electric motor which can be seen at the top left-hand side of the drawing. The normal reproducing diaphragm has been removed from the ring-shaped carriage at the top of the machine and replaced by a special measuring lever system. A sapphire reproducing stylus is connected by a vertical rod to the horizontal lever C with its fulcrum at C^{III}. A backwards extension of the lever (out to the right) is provided so that a counterweight can be added to balance the lever precisely. The lever C is 205 mm in length and its end rests upon a stiff horizontal wire (d) connected to a second lever (ef) which in turn transfers its motion through wire q to a fine piece of brass strip (g′) holding a glass siphon tube (m). This strip is suspended by the vertical torsion wire (h) and the whole movement is kept together by the action of the hair spring (i). The siphon transfers ink from the well (n) on to a strip of paper (O^{III}) which is driven forward by its own battery and motor (O^{II}).

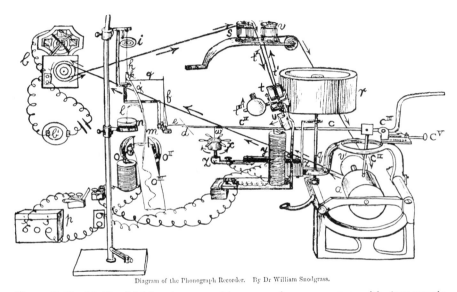

Diagram of the Phonograph Recorder.   By Dr William Snodgrass.

Figure 2.62 McKendrick's phonographic measuring apparatus with 'automatic finger and thumb' [140]. Courtesy: Royal Society of Edinburgh.

The paper transport mechanism was arranged so as to produce a slight up-and-down motion to the paper as well as moving it forward so that the trace was made as a series of dots. The relative speeds of the motors were such that 20 feet of recording paper were used for one revolution of the phonograph cylinder, representing some $7\frac{7}{8}$ inches of recorded track. Otherwise expressed, one foot of paper represented $\frac{1}{40}$th second of real time, being a magnification of approximately 30 times. The lever system magnified the indentation by about 1,000 times. The fact that the lever C rested on the horizontal wire meant that it could accommodate the horizontal tracking of the phonograph reproducer carriage along the helix of the record, but this feature in itself created some other difficulties. When the lever was rising it could, in certain circumstances, part company with the wire (d) altogether.

It was also difficult for the system to work properly for both quiet, shallowly recorded sounds and for loud, deeply indented sounds without fine adjustment of the stylus position. An automatic system was devised to alleviate this problem. A threaded vertical rod (q) which controlled the height of the ring carriage had mounted on it the drum (r), the surface of which was covered with sandpaper. The drive cord from the motor to the phonograph was taken over a fixed pulley wheel (S), was twisted over a vertical pulley (t), ran back to the pulley (v) and hence to the phonograph. The result of this configuration was that the drive cord passed twice alongside the sandpapered drum, once going down and once going up. If the pulley (t) were to be tilted one way, the cord would rotate the drum causing the screw (q) to turn and the carriage carrying the lever to rise. If the tilt of the pulley was the other way, the other part of the cord would come into play and the drum would rotate in the opposite direction causing the carriage to fall. It now only remains to explain how the tilting of the pulley was achieved. A fine platinum wire (w) attached to the lever c was dipped into a fixed cup of mercury (x). If the lever was too high, the cord which caused lowering was brought into play, but once contact was established between the wire and the mercury the electromagnet (y) was energised and the other cord was pressed against the drum. It was, in fact, a simple up/down servo system which did away with the need for constant manual adjustment of the screwed rod q. The author rather delightfully refers to it as the 'automatic finger and thumb'!

An earlier instrument described by McKendrick [141] seems to have been a somewhat less complicated affair with a simple linkage between the stylus and a lever making marks on a rotating drum, with manual adjustment to compensate for the lateral movement of the stylus along the phonograph record. In this simpler machine the speed of rotation of the cylinder when making tracings varied between one revolution in five minutes and one revolution per hour. The lever system gave a magnification of 70 times. The two papers which have been quoted appear to describe a total of three different machines altogether. The simple version just described was originally driven from a water paddle wheel through eight different belt and pulley linkages. In the author's words 'This apparatus which was large and clumsy was discarded for the smaller and more convenient apparatus consisting of a train of four sets of wheels and pinions, so geared as to obtain a gradually diminishing rate of motion'.

Sounds of various types were recorded and traced with these machines, including that produced by Mr Alfred Graham's apparatus [142] shown in figure 2.63. Here we have a telephone receiver acoustically coupled through a flexible tube to a carbon granule microphone connected back again to the receiver. This is an early example of the use of a feedback audio-oscillator,

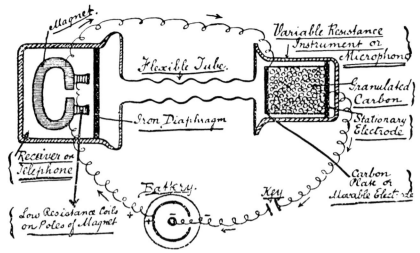

Figure 2.63 Graham's electrical oscillator [142]. Courtesy: Royal Society of Edinburgh.

although not the first; credit for that must go to D E Hughes [143,144]. McKendrick's account of this oscillator tells us that the pitch, loudness and quality of the sound produced could be varied by changing the strength of the battery, changing its polarity or varying the length and pressure of the air in the flexible tube. He also suggested that it might form the basis of a new type of electrical musical instrument:

> The diaphragms might be tuned to the notes of the scale, and by pressing on keys and thus completing the circuit musical notes having something of the quality of those brass instruments might be produced. Possibly also by piercing holes at the proper distances in the flexible tube these holes might be fingered so as to produce different sounds and we might thus have an electric flute.

E W Scripture, in the USA [145,146] constructed a particularly well engineered piece of apparatus for making similar measurements from flat-disc records with side-to-side modulation of the grooves—see figure 2.64(*a*). The electric motor seen at the right-hand side of the diagram drives a brass drum (labelled 'far drum') through a set of reduction gears. This drum's rotation is transmitted to a second drum (the 'near drum') by a long belt of paper, gummed carefully at the join. This paper belt is also smoked to receive the trace of the waveform. The gramophone record is placed on the turntable seen at the left of the figure (also seen reversed left to right in the enlarged view of figure 2.64(*b*)). The turntable is rotated by means of a belt from the near drum, and a further pulley system labelled 'pulley for side movement' drives a screw mechanism which moves the whole turntable assembly sideways so that the groove of the record always remains precisely under the stylus. This stylus is supported by a gimbal joint held by an arm fixed solidly to the main body of the machine and it is also connected to a long light lever, at the far end of which is a point which traces out the waveform on the smoked surface of the belt. Several pins projecting from the bottom of the turntable complete the circuit of a second solenoid-operated stylus (not shown here) which makes marks on the paper at intervals, to aid in the location and recognition of particular sections of the recording.

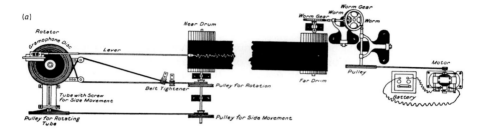

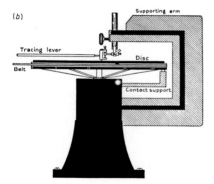

Figure 2.64 Method of amplifying and making traces of groove modulations on flat gramophone records—developed by Scripture:
(*a*), plan view of whole apparatus;
(*b*), side elevation of turntable (note: reversed left to right from view in (*a*)) [145].
Courtesy: Carnegie Institute.

To judge from the detailed account given in Scripture's paper, the whole thing was constructed with meticulous attention to detail. For example, the tracing arm was made out of 4 mm diameter Japanese reeds near the fulcrum, the rest being 2 mm German straw—apart from the very tip for which 1 mm French straw was used. The tracing point itself was a rounded glass rod, carefully hinged to the lever and held at just the precise angle to the paper for optimum results. The account even goes so far as to give the addresses from which the various materials could be obtained; the German straw from Leipzig, the French from Paris and the reeds by stripping down brushes which could be bought in the USA. Oddly enough, although the account of the instrument is enormously detailed the actual length of the lever arm is not given. However the magnification is stated (see later) and Scripture states that he used drums which were situated 100 ft apart, giving a paper band twice that length.

Details are also given as to how best to produce the carbon film on the paper belt, and since quite a number of the devices mentioned in this chapter have relied on this recording medium, it may be of some interest to quote the relevant passage.

> The smoking may be done by an ordinary gas flame provided it is sufficiently rich in carbon. A tube with a series of holes making a set of small flames is convenient, or a set of wax tapers may be used. Flames that deposit sticky smoke are to be avoided. When a record is to be fixed, a solution of shellac is spread by a brush on the back of the paper before it is removed from the drums; this fastens the smoke from the back evenly to the paper and produces a dull surface.

Suggestions are also given for mounting the whole apparatus in order to avoid vibrations, for the movement of the stylus was obviously very light and delicate and would have been upset by the slightest disturbance.

> The best method to avoid jarring is to mount the apparatus on two tables standing on a cement floor. When this cannot be done, the tracing portion may be mounted on a platform or table suspended by ropes and springs; the jarring of the floor is not transmitted to the apparatus; the jarring of the ceiling may however cause trouble. The method of resting the platform on rubber cushions that may be blown up is suggested. The far drum and the countershafts are placed on a table as far as space permits from the tracing portion (hence the desirability of the long paper belt); the motor may be placed on this table or on the floor. The motor should run with the minimum of jarring; if necessary it may be placed in a box of sand.

The lever used in this apparatus provided magnification of the groove modulation of up to 300 times, and the timescale of the traces was such that 1 mm represented 0.4 ms of recorded time. An example of the type of trace produced is shown in figure 2.65. This shows the waveform produced by an orchestra, and a fair amount of detail is obviously visible. (The long trace has been cut into sections and pasted up to provide a display of convenient shape.)

A further model of the apparatus was also constructed—see figure 2.66. This made use of a compound double-lever system, introduced with the intention of providing even greater magnification. However, even with the most careful construction and adjustment, the play in the various joints

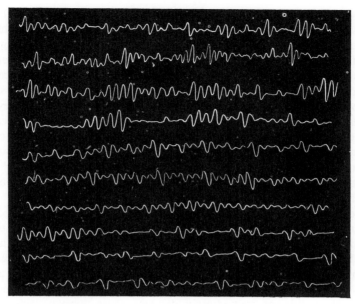

Figure 2.65 Waveform of an orchestra as inscribed by Scripture's apparatus [145]. Courtesy: Carnegie Institute.

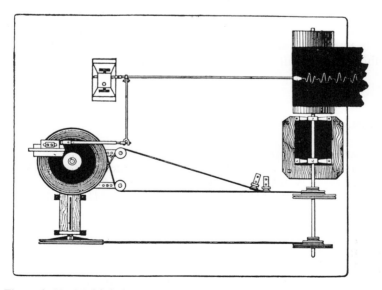

Figure 2.66 Multiple-lever version of Scripture's apparatus [145]. Courtesy: Carnegie Institute.

limited the useful magnification to only about 125 times, and on the whole the author preferred to stick to the simpler single-lever model. The actual measurement of waveforms with these instruments was a very long drawn-out affair, a rotation of the record once in five hours being recommended! In Scripture's admirably detailed description there are two sentences which jar upon the sensibilities of the modern reader. Describing the sideways motion of the turntable he refers to the belt from the second pulley 'which shoves the disc sideways'. Again, describing the precautions necessary to ensure that the record was accurately centred on its shaft—'It is carefully shoved until the point takes the middle position'. One can only assume that in American English usage of the early 1900s the verb 'to shove' did not have quite the connotation of roughness and lack of care which we associate with it.

The meticulous work of E W Scripture is perhaps a fitting end to this present chapter, for nothing could better illustrate the lengths to which experimenters had to go to secure reasonably accurate representations of speech sounds. As we have seen, work on the development of these direct diaphragm methods continued well into the twentieth century, and although more advanced means of waveform tracing had been developed in the 1890s, they did not immediately displace these simpler techniques for acoustic work. In his *Handbuch der Physik* published in 1909 [147], for example, Winkelmann was still discussing the phonautograph, the phonographic method and many of the other devices which have been described in this chapter. In electrical engineering textbooks of the same period however, the diaphragm method, which was largely experimental in nature and which never proved to be very satisfactory for alternating current work, had been almost entirely superseded by the commercially more successful methods to be dealt with in the following chapters.

References

[1]   Wheatstone C 1834 *Phil. Trans. R. Soc.* **124** 583–591
[2]   Bowers B 1975 *Sir Charles Wheatstone* (London: HMSO) ch 5
[3]   Thomson W 1853 *Phil. Mag.* (Series 4) **5** 393–405
[4]   Feddersen B W 1861 *Pogg. Ann. Phys. Chem.* **113** 437–67
[5]   Feddersen B W 1862 *Pogg. Ann. Phys. Chem.* **116** 132–71
[6]   Report 1864 *Electrician* (Series 1) **5** 111
[7]   Lorenz L 1879 *Wied. Ann. Phys. Chemie* **7** 161–93
[8]   Rood O N 1869 *Am. J. Sci.* (Series 2) **48** 153–63
[9]   Rood O N 1871 *Am. J. Sci.* (Series 3) **2** 160–7
[10]   Trowbridge J and Sabine W C 1889/90 *Proc. Am. Acad. Arts Sciences* (New series) **17** (Whole series) **25** 109–23
[11]   Trowbridge J 1890 *Phil. Mag.* (Series 5) **30** 323–35

[12] Report 1890 *Electrician* **26** 65
[13] Trowbridge J 1891 *Am. J. Sci.* (Series 3) **42** 223–30
[14] Trowbridge J 1894 *Phil. Mag.* (Series 5) **38** 182–8
[15] Fleming J A 1916 *Principles of Electric Wave Telegraphy and Telephony* 3rd edn (London: Longmans, Green) pp32–3
[16] Décombe L 1898 *L'Éclairage Électrique* **14** 569–71
[17] Battelli A and Magri L 1903 *Phil. Mag.* (Series 6) **5** 1–34, 620
[18] Marchant E W 1900 *Nature* **62** No 1609, 413
[19] Zehnder L 1902 *Ann. Phys.* (Series 4) **9** 899–918
[20] Report 1902 *Electrical World and Engineer* **40** 1049
[21] Report 1902 *Electrician* **50** 257
[22] Boys C V 1890 *Proc. Phys. Soc.* **11** 1–15
[23] Boys C V 1890 *Phil. Mag.* **30** 248–60
[24] Schuster A and Hemsalech G 1899 *Phil. Trans. R. Soc.* **139** 189–213
[25] Hehl N 1902 *Phys. Z.* **3** 547–552
[26] *See also* 1903 *Science Abstracts A* **6** No 563 204
[27] Wilson H A 1902 *Phil. Mag* (Series 6) **4** 608–14
[28] Gehrke E 1905 *Z. Instrumen.* **25** 33–7, 278–80
[29] Ruhmer E 1905 *Elektrotech. Z.* **26** 143–5
[30] Report 1905 *L'Éclairage Électrique* (Suppl.) **43** LXIX
[31] Editorial 1906 *Electrician* **56** 868 (*see also* 881)
[32] Gehrke E 1906 *Electrician* **56** 1020–1
[33] Gehrke E *see* [28]
[34] Report 1905 *L'Éclairage Électrique* **45** 320
[35] Report 1905 *Electrician* **55** 981
[36] Fleming J A *see* [15] pp40ff
[37] Diesselhorst H 1907 *Science Abstracts A* No 1895 **10** 620
[38] Fleming J A *see* [36]
[39] Paalzow A 1861 *Pogg. Ann.* **112** 567–87
[40] Paalzow A 1863 *Pogg. Ann.* **118** 178–82
[41] *see* [20, 21]
[42] Phillips V J 1980 *Early Radio Wave Detectors* (London: Peter Peregrinus) p7
[43] Dowsett H M 1920 *Wireless Telegraphy and Telephony* (London: Wireless Press) pp234ff *see also* 1923 *Dictionary of Applied Physics* vol 2, ed Sir Richard Glazebrook (London: Macmillan) p631
[44] Drysdale C V and Jolley A C 1924 *Electrical Measuring Instruments: Part II* (London: Benn) pp327ff
[45] König R 1864 *Pogg. Ann.* **122** 242–5
[46] König R 1873 *Phil. Mag.* (Series 4) **45** 1–18, 105–14
[47] Winkelmann A 1909 *Handbuch der Physik* vol 2 (*Akustik*) (Leipzig: Barth) pp166ff
[48] Poynting J H and Thomson J J 1900 *A Textbook of Physics: Sound* 2nd edn (London: Griffin) p47
[49] König R 1882 *Quelques Expériences d'Acoustique* (Paris: Lahure) ch 7
[50] Gavarrett J 1877 *Phénomènes Physique de la Phonation* (Paris: Masson) pp393ff
[51] Frölich O 1887 *Elektrotech. Z.* **8** 210–17
[52] Brown J G 1911 *Phys. Rev.* **33** 442–6

[53] Bouasse H 1926 *Acoustique Général: (Ondes Aériennes)* (Paris: Delagrave) Section 204ff (note chapter wrongly headed III)
[54] Coe B 1977 *The Birth of Photography* (London: Book Club Associates) ch 5
[55] Dowsett H M *see* [43] p237
[56] Merritt E 1893 *Phys. Rev.* **1** 166–76
[57] Nichols E L and Merritt E 1898 *Phys. Rev.* **7** 93–101
[58] Nichols E L 1894 *Proc. Am. Assoc. for the Advancement of Science, 42nd Meeting (Madison, Wisconsin) August 1893* (Salem) pp57–71
[59] Doumer E 1886 *C. R. Acad. Sci., Paris* 1886 **103** 340–2 (*see also* 1887 **105** 222–4)
[60] Hallock W 1895 *Phys. Rev.* **2** 305–7
[61] Marbe K 1906 *Phys. Z.* **7** 543–6
[62] Bouasse H *see* [53] p327
[63] Brown J G *see* [52]
[64] Bouasse H *see* [53] p313
[65] Austin L W 1901 *Phys. Rev.* **12** 121–4
[66] Weber R 1901 *Ann. Phys.* (Series 3) **6** 565–9
[67] Robinson L T 1905 *Trans. Am. IEE* **24** 185–214 (extensive quote pp189–190)
[68] Colley R 1885 *Wied. Ann.* **26** 432–56
[69] Blake G G 1928 *History of Radio Telegraphy and Telephony* (London: Chapman and Hall) pp208–9
[70] Phillips V J *see* [42] p191
[71] Töpler A 1866 *Pogg. Ann.* **128** 126–39
[72] Hervert J 1872 *Pogg. Ann.* **147** 590–604
[73] Bouasse H *see* [53] p328
[74] Winkelmann A *see* [47] p169
[75] König R *see* [46] second part
[76] König R *see* [49]
[77] Deschanel A P 1882 *Elementary Treatise on Natural Philosophy* 6th edn (Glasgow: Blackie) pp936ff
[78] Atkinson E 1893 *Elementary Treatise on Physics* (a translation of Ganot's *Éléments de Physique*) (London: Longmans, Green) Section 256, p238
[79] Lommel E 1889 *Experimental Physics* (London: Kegan, Paul, Trench and Trubner) p487
[80] Hallock W *see* [60]
[81] Hallock W 1896 *Am. Ann. of Photography* p21
[82] Winkelmann A *see* [47] p154
[83] Catchpool E 1900 *The Tutorial Physics* vol 1, *Textbook of Sound* 3rd edn (London: W B Clive, University Tutorial Press) pp185ff
[84] Stein S Th 1876 *Pogg. Ann.* **159** 142–8
[85] Miller D C 1916 *The Science of Musical Sounds* (New York: Macmillan) pp70ff
[86] Atkinson E *see* [78] Section 245 p228
[87] Gavarret J *see* [50] pp353ff
[88] Winkelmann A *see* [47] p155
[89] Read O and Welch W L 1959 *From Tinfoil to Stereo* (Indianapolis: Sams) p5
[90] Catchpool E *see* [83]

[91] Moigno F 1859 *Cosmos* (*Revue Encyclopédique Hebdomadaire des Progrès des Sciences*) (Series 1) **14** 314–20
[92] Scott É-L 1861 *C.R. Acad. Sci.*, *Paris* **53** 108–11
[93] Bell A G 1877 *J. Soc. Telegr. Eng.* **6** pp385–416
[94] Du Moncel Comte Th A L 1882 *Le Microphone, Le Radiophone et Le Phonograph* (Paris: Librairie Hachette) pp37ff
[95] Du Moncel Comte Th A L 1879 *The Telephone, The Microphone and The Phonograph* (New York: Harper) pp39ff
[96] Prescott G B 1878 *The Speaking Telephone, Talking Phonograph and Other Novelties* (New York: Appleton) pp67ff
[97] Blake G G *see* [69] p13
[98] Blake C J 1876 *Arch. of Opthalm. and Otology* **5** 108–113
[99] Blake C J 1878 *Arch. of Opthalm. and Otology* **7** 457–64
[100] Barlow W H 1874 *Pop. Sci. Rev.* **13** 278–88
[101] Barlow W H 1873–4 *Proc. R. Soc.* **22** 277–89
[102] Blake E W 1878 *Nature* **18** 338–40
[103] Blake E W 1878 *Am. J. Sci.* **16** 54–9
[104] Frölich O 1887 *La Lumière Électrique* **25** 180–7
[105] Frölich O 1891 *Electrician* **28** 59–61
[106] Frölich O 1887 *Elektrotech. Z.* **8** 210–17
[107] Frölich O 1889 *Elektrotech. Z.* **10** 345–8, 369–76
[108] Feldmann C P 1894 *Wirkungsweise, Prüfung und Berechnung der Wechselstrom Transformatoren* (Leipzig: Oskar Leiner) p395
[109] Duddell W B 1897 *Electrician* **39** 636–8
[110] Blondel A 1891 *La Lumière Électrique* **41** 401–8
[111] Barr J M *et al* 1895 *Electrician* **35** 719–21
[112] Blondel A 1894 *La Lumière Électrique* **51** 172–5
[113] Robinson L T *see* [67]
[114] Thomson E 1888 *Electrician* **20** 350–3
[115] Thomson E 1888 *Telegraphic Journal and Electrical Review* (*London*) **22** pp108–11
[116] Swinburn J 1888 *Practical Electrical Measurements* (London: Alabaster/Gatehouse) and (New York: Van Nostrand) pp150–1
[117] Guyau M 1913 *C. R. Acad. Sci.*, *Paris* **156** 777–9
[118] Miller D C *see* [85]
[119] Miller D C 1937 *Sound Waves, their Shape and Speed* (New York: Macmillan) pp12ff
[120] Shrader J E 1937 *Physics for Students of Applied Science* (New York: McGraw-Hill) p501
[121] Twyman F and Dowell J H 1922 *J. Sci. Instrum.* (preliminary volume, usually bound in with volume 1) pp12–4
[122] Bailey A E 1983 *J. Phys. E: Sci. Instrum.* **16** 937–4
[123] Boeke J D 1895 *De Natuur* pp15–22 (This journal does not seem to have used volume numbers.)
[124] De Vries L 1971 *Victorian Inventions* (London: John Murray) pp180–1
[125] Guerot A 1881 *La Lumière Électrique* **3** 359–60
[126] Mayer A M 1877–8 *Nature* **17** 469–71

[127] Du Moncel Comte Th A L *see* [94] and [95]
[128] Fleeming Jenkin and Ewing J A 1875–8 *Proc. R. Soc. Edinburgh* **9** pp579–581, 583
[129] Fleeming Jenkin and Ewing J A 1879 *Trans. R. Soc. Edinburgh* **28** 745–79
[130] Fleeming Jenkin and Ewing J A 1878 *Nature* **18** 340–3, 394–7, 454–6
[131] Ellis A J 1878 *Nature* **18** pp38–9
[132] Boeke J D 1891 *Pflügers Archiv für Phisiologie* **50** 297–318
[133] McKendrick J G 1896 *Trans. R. Soc. Edinburgh* **38** 765–85
[134] Rollefson C J 1895 *Phys. Rev.* **2** pp141–3
[135] Rollesson C (name misprinted) 1894 *L'Éclairage Électrique* **1** 276, 1895 **2** 461
[136] Hermann L 1889 *Pflüger's Archiv für Physiologie* **45** 582–92
[137] Bevier L 1900 *Phys. Rev.* **10** 193–201
[138] Bevier L 1902 *Phys. Rev.* **14** 171–181
[139] Bevier L 1902 *Phys. Rev* **14** 214–220, 1902 **15** 44–50, 271–275, 1905 **21** 80–9
[140] McKendrick J G 1895–7 *Proc. R. Soc. Edinburgh* **21** pp194–205
[141] McKendrick J G *see* [133}
[142] McKendrick J G 1895–7 *Proc. R. Soc. Edinburgh* **21** 46–8
[143] Kennelly A E 1923 *Electrical Vibration Instruments* ch 23 (New York: McMillan)
[144] Hughes D E 1883 *J. Soc. Telegr. Engineers and Electricians* **12** 245–50
[145] Scripture E W 1906 *Researches in Experimental Phonetics—the Study of Speech Curves* (Washington: Carnegie Institution) ch 2
[146] Miller D C *see* [85] pp77–8
[147] Winkelmann A *see* [47]

3

Contact Methods

Summary

This chapter will describe the contact or 'point-to-point' method which was the mainstay of electrical waveform determination from 1880 up to the introduction of the mechanical oscillographs at the end of the century. It may be regarded in principle as being an electrical version of the simple stroboscope. The stroboscope itself was actually used in its ordinary optical form to make measurements on vibrating strings etc. and also to make observations of electric arc lights.

The introduction of the electrical version, by Joubert in Europe and Thomas in the USA, will be outlined and the probable use of the method at an earlier date by Lenz and Wheatstone will be considered.

In its earliest form, it was an extremely tedious technique to use. After a description of these simple 'manual' systems, the later automatic versions will be considered. Once fitted to machines, these contacts were available for use in applications other than waveform observation, and some of these will be mentioned. The chapter will conclude with a consideration of the extension of the method for the examination of repeated transient waveforms, and the 'rheotomes' used by electrophysiologists will be described.

3.1 The stroboscope

At the time when the need for an effective method of observation was being most acutely felt with the introduction of alternating current systems, the principle of the stroboscope was already well known. This had been developed by Töpler[1] in the 1860s, and he had used his clockwork-driven pierced disc for the study of vibrating strings, tuning forks and the like. Indeed, reference to the paper describing his invention (which he called the 'vibroscope'), written in the Philosophical Magazine indicates that the principle was not entirely new even at that time, for he refers to its earlier use by Professor Magnus to study the behaviour of droplets in water jets.

Simple stroboscopic discs were later used by Hospitalier [2,3] and by Beckit Burnie [4,5] to study flicker in the emission of light from electric arc lamps driven from alternating sources. Hospitalier's apparatus, dating from 1901, which he called the 'arcoscope' simply allowed the light from the arc to be interrupted by a slotted disc driven by an induction motor running from the same AC supply as the arc itself. The image of the arc was cast upon a screen and by allowing the induction motor to slip slightly in speed the flicker of the arc could be slowed down and studied at leisure.

W Beckit Burnie's slightly earlier apparatus (1897) can be seen in figure 3.1. A system of mirrors mounted on a swinging arm allowed the arc to be viewed from any angle, the light beam ending up horizontal as shown in figure 3.1(*a*). The stroboscopic disc was mounted on the shaft of a synchronous motor and could be set at any one of 32 angular positions by means of the simple detent mechanism shown in figure 3.1(*b*). The light from the arc was interrupted by the disc and allowed to fall on a photometer which measured the light output at the same point in the electrical cycle in each revolution of the disc. A bolometer could also be used to study the heat emission from the arc.

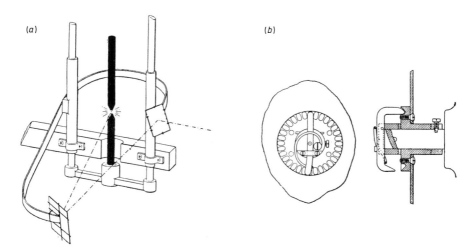

(*a*)   (*b*)

Figure 3.1 Beckit Burnie's apparatus [4] for viewing an electric arc through a stroboscopic disc: (*a*), arrangement of mirrors; (*b*), detent mechanism for setting the disc at different angular positions on the motor shaft.

A modification of the stroboscopic principle was also employed in the ingenious piece of apparatus invented by Mikola [6] in 1906 for the purpose of observing vibrating strings. This is shown in figure 3.2. The vertical drum K is driven by a variable-speed motor M. Its surface is totally black save for one vertical white strip. The image of the string whose motion was to be observed was cast upon the drum by the lens system P. If the string were stationary the appearance of the drum would be a dark gray blur due to the persistence of vision, with a black horizontal line representing the image of the string where no light at all fell on the drum. When the string was set into vibration, successive positions of the drum would correspond to different positions of the string, and a wavy black line would then be seen. If the rotation of the drum were synchronised to the string frequency a sinusoidal trace of the sort seen in figure 3.3 could be produced. The idea could be extended somewhat by having several vertical white lines on the drum, and if arranged so that successive images would coincide, a brighter picture with greater contrast could be obtained.

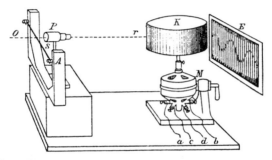

Figure 3.2 Mikola's rotating drum for the observation of vibrating strings [6].

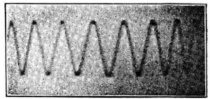

Figure 3.3 Image produced by Mikola's drum (Winkelmann 1909 *Handbuch der Physik* vol 2 (Barth: Leipzig) p171).

3.2 Simple manually set contacts

The electrical version of the stroboscopic method, which was widely used after its introduction by the French physicist Jules Joubert in 1880, was known as the 'contact' method, or the 'point-to-point' method. The principle of operation may be understood by reference to figure 3.4. An alternator (not shown) connected to a and b supplies the arc lamp seen at the bottom of the diagram, and the voltage waveform is to be determined.

Mounted on the shaft of the alternator itself is the drum D which is made of two discs. The nearer disc is made of ebonite; the other one is made of brass. A small tongue of brass f projects along the edge of the insulating disc in such a manner that the two springy metal brush contacts are connected together momentarily once in each rotation of the shaft.

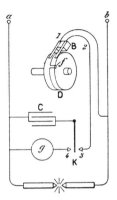

Figure 3.4   Joubert's rotating-contact method of waveform plotting [14].

To use the device, the key K is first pressed against contact 3 so that the capacitor is topped up with charge at precisely the same point in each cycle and attains the voltage present in the circuit at that time. The key is then released and the voltage on the capacitor is measured by some means. One method, that shown in figure 3.4, is to use a ballistic galvanometer, its 'throw' being a measure of the charge stored, and hence of the voltage. An ordinary moving-coil voltmeter of high resistance could also be used to measure the voltage directly, and the quadrant electrometer was another type of meter which was employed for this purpose. Having measured the capacitor voltage, the next step is to rotate the angular position of the drum on the shaft so that contact will be made at a different point in the cycle, and this procedure is then repeated until the whole of the waveform has been adequately delineated 'point by point'. This was, of course, a sampling method—a mechanical forerunner of the sampling oscilloscope of more recent times.

As previously mentioned, it was Joubert, sometime General Secretary of the Société Français de Physique [7,8], who introduced the method in 1880, and in fact it was often referred to simply as 'Joubert's method'. There is, however, evidence to show that it had been used previously by other experimenters. In 1849, the Russian H F E Lenz wrote a paper describing his experiments with a primitive magneto type of electrical generator [9]. (This is the same Lenz whose name is forever associated with the law of electrical action and reaction which he propounded in

1833/4.) In the course of these experiments he observed that for maximum voltage output from the machine, the commutator contact which transmitted the voltage from the rotating coils had to be set in a particular angular position. He seems to have realised that this idea could be extended so that by setting the contact at various angles a complete picture of the output waveform could be built up. The second part of his paper published in 1854 [10] has a diagram of a commutating contact which can be set at various angles on the generator shaft and also contains several examples of curves of voltage obtained in this way. Lenz's contact is shown in figure 3.5. The wooden cylinder GHJK was fixed on the machine axle by the screw FE. At each end of the cylinder is an iron disc. One disc, CD is whole. The other disc AB has six sectors cut out and then fixed back in, but insulated from the rest of the disc. Only the parts marked $\alpha\beta$, $\alpha'\beta'$, etc are in metallic contact with the centre. The coil in which the voltage is induced is connected to the commutator, so that the brushes make contact for $3°$ out of $60°$ or one twentieth of the cycle. Oddly enough, the paper shows a four-pole machine, but he says in the text that the contact was used with a 'Storer' machine. There is one of these machines in the Science Museum in South Kensington, and it has six poles—which ties up with the number of contacts on the commutator.

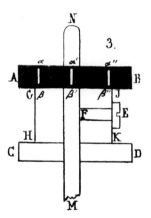

Figure 3.5 Lenz's rotating contact [10].

Brian Bowers, in his book on the life of Charles Wheatstone [11] describes a small weight-driven magneto, also in the Science Museum collection, which is fitted with a contact disc to allow a series of point-to-point measurements to be made. Indeed, Dr Bowers shows several waveforms measured by himself with the aid of this disc. This machine dates from 1858, and since Charles Wheatstone understood German, and since he never claimed any credit for this method of waveform observation, we may conclude that he was aware of Lenz's work when he constructed this machine.

In 1891, a Mr M E Thompson presented the results of some measurements on an open-coil dynamo to the American Institution of Electrical Engineers. During the discussion which followed [12], Dr Geyer remarked that the Joubert contact method which the author had used had been employed in a similar investigation undertaken by Professor Benjamin Franklin Thomas as early as 1879 or 1880, (see also [29]). Since this was a date before the publication of Joubert's paper, the opinion was expressed that Professor Thomas should be approached to ascertain the facts of the matter, and possibly to establish a prior American claim to the method. In June 1892, Thomas duly presented a paper before the Institution entitled 'Notes on contact methods for current and potential measurements' [13] in which he confirmed that he had indeed devised a method of this type in 1880, but that since he had shortly afterwards moved to another post, he had no opportunity to resume its use until 1885. The 1880 work, though, had been presented to the American Association for the Advancement of Science at its Boston meeting held in August 1880. This would appear to be a classic instance of the same idea occurring independently to two people at the same time. Which of the two, Joubert or Thomas, was actually *first* is difficult to establish, but in view of the prior work of Lenz and Wheatstone the matter is really of little importance.

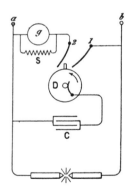

Figure 3.6   Blondel's version of the rotating contact [14].

There was a variant of the Joubert method, sometimes called the 'double-contact' method, usually ascribed to André Blondel [14–17] and illustrated in figure 3.6. Here the rotating contact first touches the springy metal brush 1 which charges up the capacitor to the voltage on the line at that time. It then leaves this contact and discharges the capacitor via brush 2 and through the galvanometer g. Provided the capacitor is completely discharged during each revolution, the galvanometer will settle at a steady reading proportional to the voltage. The whole thing is easily calibrated by applying known voltages to ab.

Several important points need to be stressed about the Joubert method. Firstly, it was a very tedious and repetitive job to measure a sufficient number of points to delineate the waveform accurately. Secondly, since it took so long, there was no guarantee that the waveform had not changed during the time of measurement. This was particularly relevant if, for example, one were trying to record the voltage waveform on a public supply line, with consumers switching their apparatus on and off during the course of the measurements. Finally, it was a method which was only suitable for repetitive waveforms; it could not be used for transient phenomena (unless the phenomenon could be arranged to repeat at regular intervals—see later). The situation was well summarised by an anonymous author in *Nature* in 1900 [18]:

> This method is open to two objections. In the first place, it is only applicable to cases in which the waveform is undoubtedly steady, all transient effects being obviously unobservable by such a process, and, secondly, it is so lengthy that elaborate researches are practically excluded. As much as four or five hours may, indeed, be spent in obtaining a single curve, and then, even after all this labour, it is more than possible that the conditions will be found to have altered during the experiment, and the curve, in consequence, to be useless.

In its simplest and crudest form (as on the Lenz and Wheatstone generators) the machine had to be stopped between readings, in order to move the contact to a new position on the shaft. This nuisance was soon obviated by the use of a brush mounted on an arm which could be swung in an arc around the contact disc as in the typical arrangement dating from 1890 shown in figure 3.7 [19–22]. Here the springy brush A can be moved into various positions around the whirling contact B and a calibrated scale is provided so that the angular setting can be noted. Another contact system of similar type and dating from about the same time is shown in figure 3.8 [23,24]. The large cylinder is a belt pulley of the usual sort mounted on the axle of the machine, while the smaller cylinder is made of hard rubber and has a brass contact bar set into its surface. The swinging spring contact is attached to the disc on the right, the angular position of which can be altered by means of the handle. The periphery of the disc has a scale which, used in conjunction with the vertical pointer mounted on the frame of the machine (see figure 3.8(*b*)), enables the operator to read the contact position.

One of the factors governing the accuracy of the measurement was the precision with which the angular position of the brush or spring contact could be set. A Mr Brackett, writing to the *Electrical World and Engineer* in 1906, proposed the use of a double Prony brake to improve the accuracy of setting [25]. A Prony brake, (named after Gaspard Clair François Marie

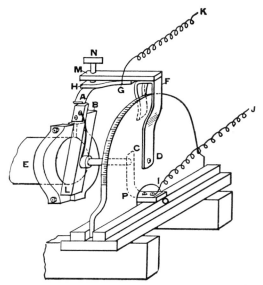

Figure 3.7   A typical arrangement of a contact mounted on a swinging arm [19].

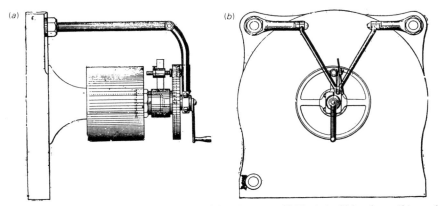

Figure 3.8   Swinging-arm contact used by Duncan [24]. (*a*), Side view; (*b*), end view.

Riche de Prony, the Chief Engineer of Bridges and Roads in Napoleonic France) consists of two shaped blocks of wood clamped around a shaft by means of bolts as in figure 3.9, being restrained from rotating by the lever resting against the stop S_1. The proposed device had two of these brakes B_1 and B_2 mounted on a secondary shaft S which carried the contact spring and around which it swivelled. The sequence of operation was as follows. B_1 was released by loosening its bolts, and its arm was moved until it hit S_2.

It was then clamped in this position. B_2 was then unclamped and B_1's arm was moved back to S_1 at which point B_2 was clamped again. Proceeding in this way, the secondary shaft could be turned through successive well defined segments of arc and the brush could thus be set at intervals around its scale in an incremental fashion.

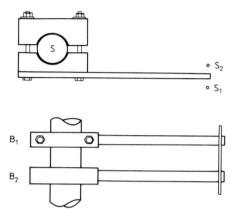

Figure 3.9 Double Prony brake system for accurate setting of the angular position of the contact shaft.

C F Smith [26] preferred to use the system shown in figure 3.10. The two insulated discs (d_1 and d_2) were mounted on a bush (shown in dotted lines) running on the metal spindle (S) the pointed end of which was inserted into a depression on the end of the shaft of the alternator whose waveform was under observation. The smaller disc (d_2) was solidly fixed to the bush but the larger disc was simply a tight fit on the bush and its angular position on the spindle could be altered by exerting sufficient force. The rim of d_2 was a continuous brass strip but d_1 had only a small inset strip of brass which rested on the rim of d_2 making contact with it. The whole assembly was made to turn by pushing the discs forward until the small pin projecting from the machine axle engaged with the hole in d_2. Two metal brushes rubbed on the rims of the discs (these are not shown in figure 3.10) and acted as contacts once per revolution in the usual way. When a voltage reading had been taken, the discs were pushed out of engagement with the axle so that they were stationary; the machine was still running, undisturbed. The position of the contact point was then altered by turning d_1 on the bush, and the whole was then caused to spin again by re-engaging it with the pin. How many fingers were lost during this operation is unknown, but a rate of making observations of one reading per minute was claimed!

An alternative system which was sometimes encountered is shown in figure 3.11. An insulating drum has set into its surface a spiral conductor,

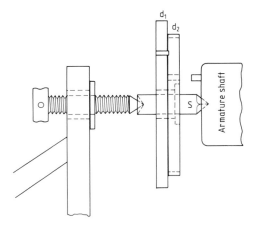

Figure 3.10 Pin clutch arrangement, allowing resetting of contact angle whilst machine shaft is still rotating [26].

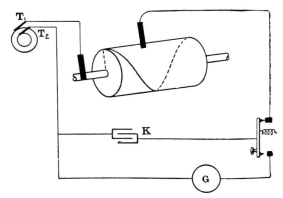

Figure 3.11 An alternative mechanism for setting the angular position of the contact [27]. Courtesy: Macmillan, New York.

the time of contact being varied by moving the stationary brush parallel to the axis of the drum [27].

It is hardly necessary to point out that brushing contacts of the kind needed for these measurements produced their fair share of problems [28]. L Searing and S V Hoffman, for example, describe the various frustrations which they encountered in the search for a satisfactory arrangement. They eventually settled for a light springy wire which lasted for six or seven hours before replacement was necessary [29]. They also described the difficulties caused by large electrostatic charges building up on the belts which drove the machines, and which were sufficient to swamp the charge they were actually trying to measure. Proper earthing overcame the problem—more or less. One attempt to overcome brush troubles is represented by the system shown in figure 3.12 [30, 31]. A fine jet of water (presumably salted or acidulated to render it conducting) issuing from the nozzle J strikes the needle as it rotates upon the axles of the alternator. The nozzle position is altered by turning the calibrated mounting disc under the clamp C. This device, dating from 1893, was reported to work quite well, but one cannot escape the feeling that it would have had other troubles and inconveniences peculiarly its own! The renowned American pioneer of radio, R A Fessenden is reported to have used a similar system, replacing the water with mercury in order to do away with problems caused by polarisation [32].

Figure 3.12 Bedell's water jet contactor [30].

Another attempt to do away with mechanical contacts altogether was made by Goldschmidt in 1902 [33–35]. His apparatus is shown in figure 3.13. The brass drum has four bars made of transformer stampings set into its surface. The transformer has the voltage waveform under observation connected to its primary coil S_1. The inset stampings represent the 'missing side' of the transformer core E so that appreciable voltage is only induced into the secondary coils S_2 when one of the bars is in the correct position. The transformer itself is mounted on a swinging arm that can be moved around the drum so as to vary the phase of the amplitude samples. The exact analysis of the operation of this arrangement is quite complicated, but in spite of certain inherent errors it seems to have given good results if the published curves are to be believed.

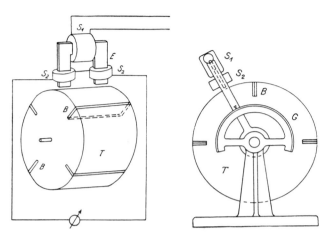

Figure 3.13   Goldschmidt's 'magnetic contactor' [35].

The provision of a 'contactless contact' was also the aim of H J Ryan who in 1899, read a paper entitled 'The determination of the wave form of alternating currents without a contact maker' before the American Institution of Electrical Engineers [36]. This he did with the aid of a specially constructed transformer which had a core of normal cross section, save for one limb which was very small in area. The result was that when an alternating voltage was applied to the primary, the flux in the small limb went from full saturation in one direction to full saturation in the other. Since the voltage in the secondary depends on the rate of change of flux and since the flux was constant except near the zero crossings of the sinusoid, the secondary voltage consisted of a series of sharp pulses occurring at those zero crossings. Ryan wished to observe the waveform of the mains voltage. He therefore applied these pulses to one set of coils of a dynamometer wattmeter and applied the mains voltage to the other. It will be recalled that this type of meter acts as a multiplier, and so the needle of the meter settled at a value proportional to the value of the waveform at the times of the pulses. By means of a subsidiary electrical machine he was able to obtain a sinusoid of variable phase relative to the mains and this was used as the input voltage to the special transformer. In this way he was able to move the phase of the pulses relative to the mains so that he could build up a picture of the waveform by a series of point-to-point readings on the meter.

In his original 1880 paper, Joubert did not actually use a capacitor for storing the sample voltage, but simply connected the contact to an ordinary metre-bridge potentiometer. At balance, when the potentiometer voltage was equal to the sample voltage appearing on the contact, a null reading was obtained on the indicating instrument. Many other workers also used this configuration—Professor B F Thomas, for example, in his early experiments already referred to [37,38]. With this particular method of measurement it was important that the actual contact should be made for as short a time as possible or the changing input voltage during contact time would cause balancing difficulties and loss of precision in the readings. In fact the contact time in Joubert's apparatus was as low as 50 μs. Professor C A Adams claimed that the type of contact shown in figure 3.14 was very successful in producing short contact times. Once in every revolution the steel pin on the rotating disc made contact with the steel spring S, and hence with the stiff strip B with which it was in contact. Almost immediately the motion of the pin lifted the spring away from the strip and contact was broken again. Unfortunately Professor Adams does not quote actual contact times [39] but merely says that the duration was so small as to be negligible even when using an alternator giving the relatively high frequency of 180 Hz.

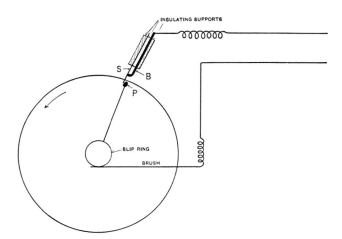

Figure 3.14 Professor Adams' method of producing contacts of short duration [39].

If the voltage being measured was rather small in magnitude, and if the contact time was short, it was sometimes found that it was difficult to find the null point using an ordinary galvanometer as an indicating instrument. R D Mershon suggested using a telephone receiver for this purpose in order to improve the sensitivity—see figure 3.15 [40–42]. In this circuit C is the rotating contact and p and q are the sliding contacts of the potentiometer. When off-balance a series of clicks can be heard in the telephone, these becoming inaudible as the circuit is adjusted to the null point. It was claimed that this rearrangement could be set to an accuracy of 1/100th of a volt. The reason for the inclusion of the capacitor K is not given: Mershon merely says that 'it may be connected in parallel with the telephone'. It would not seem to be serving any very useful purpose, unless perhaps its presence moderated the sharpness of the clicks in some way and made them less fatiguing to the ear.

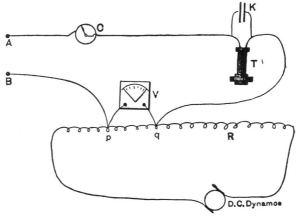

Figure 3.15   The use of a telephone receiver for extra sensitivity when measuring the voltage on the contact by means of a potentiometer circuit [40].

Having stressed the importance of a short duration of contact when no capacitor was used, it should be pointed out that when there *was* a capacitor present it could be left connected for a much longer time: it was the voltage left on the capacitor at the moment of breaking contact which was eventually measured. The provision of a short-duration contact was thus of much lesser importance in this type of system.

In most of the examples quoted so far, it was assumed that the contact-making arrangement could be mounted directly on the axle of the alternator which was producing the waveform, thereby achieving perfect synchronisation between the contact frequency and the mains frequency. Of course, it was not always possible to do this, and in such cases it was common practice to rotate the contact disc by means of a synchronous motor driven by, and locked to the supply, the waveform of which was being measured [43–45]. A typical arrangement of this type due to Fleming and dating from 1895 can be seen in figure 3.16 [46–48]. To the right can be seen the synchronous motor; to the left, the contact disc SS with the brushes mounted on the swinging arm H whose position is read on the circular scale G. Figure 3.17 shows some typical curves obtained by the use of this system. The dotted

curves were measured experimentally, point by point, and the full curve
is a flux curve obtained by integration of the voltage curve. These
synchronous-motor-driven contact discs were generally held to be less
accurate than those mounted solidly on the alternator shaft because of the
possibility of hunting or slip caused by the pressure of the contacts.

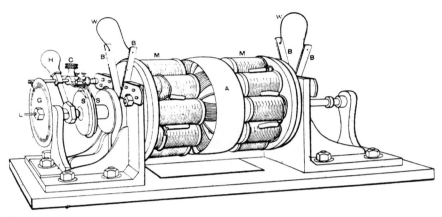

Figure 3.16 An example, constructed by Fleming, of a contact disc driven by a synchronous motor locked to the supply voltage under observation [46].

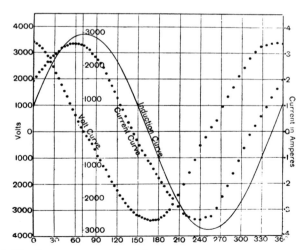

Figure 3.17 Some curves plotted by means of the synchronous motor
apparatus of figure 3.16. (The dotted curves represent the actual
measurements of voltage and current; the full curve is the flux distribution curve calculated by integration of the voltage) [46].

3.3 Semiautomatic and fully automatic contact systems and other modifications

The experimental arrangement of figure 3.18 for which a US Patent was sought in 1901 and granted in 1904 to L T Robinson, represents an attempt to speed up the measuring process and remove some of the tedium [49–51]. The synchronous motor was brought up to speed by turning the geared handle on the right and then the supply was applied so that it kept rotating at the correct speed. The contact brush 6 was fixed to the motor shaft and rotated rapidly around the stationary contact disc 7. The contact disc had four short contacts on its circumference, and four long ones; the number of contacts would depend on the number of poles in the machine. The short contacts were used to charge up a capacitor from the supply waveform, the long ones to discharge it through a mirror moving-coil galvanometer. The galvanometer deflection was proportional to the voltage at the instant of disconnection of the capacitor from the supply waveform. A photographic plate holder 15 was moved by a string and pulley system thereby providing a time axis as the contact disc was rotated manually by the lever 17. The movement of the galvanometer was simultaneously recorded by a light beam so that a voltage–time waveform was recorded on the plate. A photograph of the actual apparatus, which was made by the General Electric Company in the USA, can be seen in figure 3.19. This instrument,

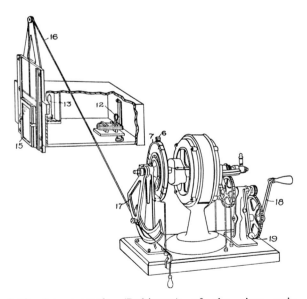

Figure 3.18 An example (Robinson) of the time axis on a photographic record obtained by a sliding plate holder linked by cord to the contact setting handle [49].

which was exhibited at the American Institution of Electrical Engineers' Conversazione in 1901, was referred to by the makers as a 'wavemeter'. Today this word is usually applied to an instrument which measures wavelength or frequency—a rather different usage, although it could be argued that if one has a means for drawing out the waveform, one also has a means of measuring its periodic time if the timescale is properly calibrated.

Figure 3.19 The General Electric wavemeter—a commercial version of Robinson's contactor [50]. Photograph: Science Museum (397/86).

T R Lyle, Professor of Physics at Melbourne [52], and F A Laws, of the Massachusetts Institute of Technology [53] also used string-operated sliding plates to produce a time axis; more will be said about their particular systems later. J M Barr and his colleagues [54–56] used a cam-operated mirror to provide the time axis deflection when recording the galvanometer movement on a photographic plate.

A further stage in automation, removing entirely the need for manual operation, was made by Blondel in 1891—interestingly enough, a decade before the Robinson apparatus [57–58]. This is shown in figure 3.20. Here there are two galvanometers and double-contact sets, one to draw the voltage waveform and one the current. The brush position is slowly moved around continuously by means of a clockwork motor, the winding key of which is visible at H. The arc lamp throws two beams of light onto the galvanometers via the mirrors M and the deflected beams fall upon a band of photographic paper driven by the same clockwork motor so that a record of the two waveforms is drawn. Figure 3.20 also shows the image of the arc lamp thrown upon a separate screen. This was to aid in keeping the luminous spot of the arc correctly adjusted relative to the mirrors as the carbons burnt down. Figure 3.21 shows an example of the voltage and current waveforms of an alternating current arc lamp drawn out by this apparatus.

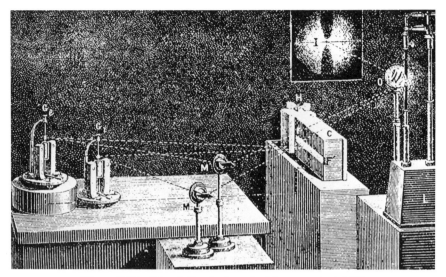

Figure 3.20   Blondel's clockwork-driven contact [57].

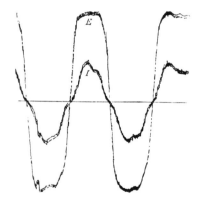

Figure 3.21   An example of some voltage and current waveforms produced by Blondel's apparatus [58].

The version of Blondel's apparatus shown in figures 3.22 and 3.23 is obviously rather less of an experimental 'rig' [59]. The drum T carrying the photographic paper p is situated inside a dark chamber and it is mounted on the same shaft as the moving contacts (r). Again, the slow movement of the contacts is provided by the geared clockwork motor H. Spare papers were stored in the cylinder marked D and the paper on the drum could be changed by inserting the arms through the velvet sleeve M on the top of the chamber.

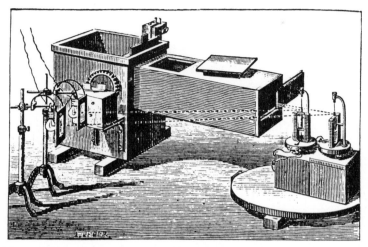

Figure 3.22 A more fully developed version of Blondel's apparatus with photographic paper transport mechanism enclosed in a light-proof chamber [59].

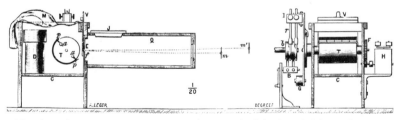

Fig. 2 et 3. — Appareil inscripteur, coupes longitudinale et transversale. C chambre noire; Q appendice; T tambour fixé sur l'arbre de l'appareil de contact; F roue dentée calée sur le même arbre et entraînée par le mouvement d'horlogerie; I tête en ébonite portant les ressorts r; G contrepoids; B balais isolés; b bagues isolées; M sac en toile noire; D boîte à papier; p pointes; a anneau; v velours arrêtant la lumière, V volet; E écran avec sa fente f; J judas; m et m' miroirs des galvanomètres.

Figure 3.23 Details of the contactor mechanism and photographic chamber of Blondel's apparatus (figure 3.22) [59].

As mentioned above, Blondel used two contact makers in order to be able to draw two waveforms simultaneously. L Duncan suggested a rather ingenious method of enabling one contact maker to serve for taking several traces [60–62]. He replaced his *permanent* magnet moving-coil galvanometers with *electro*magnet (electrodynamometer) instruments, but instead of passing a DC current through the fixed coils to produce a steady magnetic field, he passed pulses of current through the moving coils via the contactor as shown in figure 3.24 and applied the various currents and voltages to be observed to the fixed field coils. Each galvanometer showed a deflection proportional to the value of its voltage or current waveform at the instant of contact.

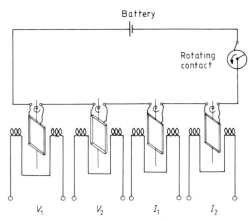

Figure 3.24 Duncan's method for enabling one contactor to be used for the simultaneous observation of several voltage and current waveforms.

R B Owens also used a dynamometer of special construction, although his stated aim was not to produce an automatic system, but a sensitive manual method of 'great range and high accuracy' [63]. His galvanometer (figure 3.25) had a moving coil suspended on two wires and was provided with an oil damping system. There were two concentric sets of fixed coils. The variable current whose waveform was to be determined was passed through the outer set. A DC current was passed through the inner set, its value being measured accurately by means of an ammeter. A pulse of current from a contactor was sent to the moving coil as in Duncan's method. If the steady direct current was set to be equal and opposite to the current in the waveform at the instant of contact, no deflection was produced. This was essentially a 'null' method, the deflection being adjusted to zero at each setting of the contact, and the current being measured accurately on the ammeter. The waveform was plotted manually, point by point.

Returning to methods of making the measurement automatically; another widely used way of advancing the contact continuously and smoothly was to gear its mounting disc to the shaft of the alternator or synchronous motor. In 1901, for example, F A Laws used a 7,200 : 1 worm reduction gearing for this purpose [64]. Figure 3.26 shows two contact

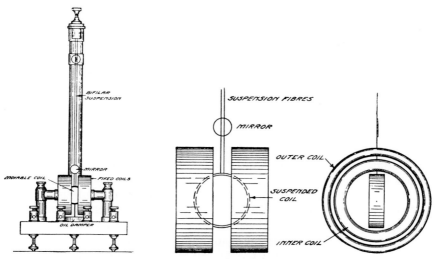

Figure 3.25 A dynamometer instrument with a double set of fixed coils used for measuring contactor voltages by a null method (Owens) [63].

makers, each consisting of an insulating disc having four brass pieces set into the circumference. The waveform across the resistor ab is to be determined. The two discs are firmly fixed to the same shaft of a synchronous motor driven from the waveform being observed. The discs are slightly staggered in phase so that contact K_2 is made first, charging up the capacitor, and immediately this has broken, K_1 discharges it through a galvanometer. The two spring contacts are rigidly fixed to each other and are mounted on a radial arm, geared to the shaft, so that they move round

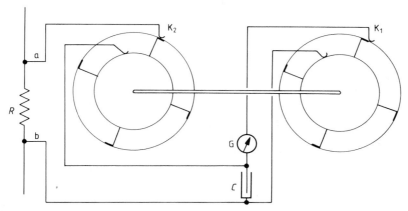

Figure 3.26 Double contact-disc charge and discharge arrangement employed by F A Laws [64].

very slowly. A wire, winding around a drum on the gearing passed over a pulley to a sliding-plate system similar to that of Robinson (figure 3.18) for photographic recording of the waveform. Others also used this method of altering the contact position by gearing from the main shaft, [65] and of linking the motion of the contact to the recording mechanism [66].

A somewhat different approach to the problem of speeding up the process of plotting the waveforms was adopted in 1897 by Professor E B Rosa, (shortly thereafter to be appointed Professor of Physics at the Wesleyan University, Middletown, Connecticut) [67–70]. His basic method of measuring the potential appearing on the contact was to use a slide-wire potentiometer as shown in figure 3.27. The alternating voltage appearing across the resistor AB was to be plotted, and this was applied to the potentiometer wire NO and the galvanometer G via the rotating contact maker CM. When the sliding contact P was in the correct position, a null reading would be obtained. The slider was linked to a recording stylus F by means of a pantograph mechanism so that the reading could be transferred directly to a piece of paper wrapped around a recording drum.

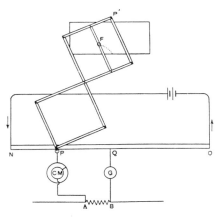

Figure 3.27   Schematic diagram of Rosa's apparatus showing the potentiometer slide P linked by a pantograph to the recording pen F. CM is the rotating contact maker; G is a galvanometer [69].

The instrument itself can be seen in figure 3.28. The user kept his eye on the galvanometer, and with his left hand turned the milled knob at the front of the apparatus which moved the slider in order to bring the potentiometer to balance. With his other hand he then pressed a key which had three effects. A solenoid was energised so that a stylus made a mark on the drum. On release of the key, a ratchet mechanism, seen at the bottom right of the apparatus, advanced the contact to the next angular position and also moved the recording drum on by one notch. With a little practice it was claimed that up to twenty points a minute could be plotted in this way. It will be noted from figure 3.27 that the electrical arrangement of the potentiometer was such as to allow it to measure voltages of both polarities.

Figure 3.28 Rosa's apparatus. The slide-wire potentiometer is on the left of the picture; the recording drum and contact advancing mechanism are on the right [69].

Rosa's curve tracer was produced commercially by the Elmer G Willyoung Co. [71], being marketed by their agent, James G Biddle of Philadelphia. Various additional accessories were also available, including a second recording solenoid which marked off the measured values of, say, the current along a strip of paper laid flat alongside the potentiometer scale. A second strip of voltage values measured in the same way could then be removed, and attached around the circumference of the recording drum. By a further procedure, involving two operators, the instrument could be used to draw out an $x–y$ type of display of the two variables. Magnetic hysteresis curves could be drawn in this manner. Another complicated procedure permitted the drawing of curves of instantaneous power as a function of time. For this purpose a second potentiometer wire was supplied parallel to the first. This was used in place of the resistor AB of figure 3.27. The first part of the operation was to tap off a section of this resistance which was proportional to the current, previously measured, at a particular instant in the cycle. The complete slide wire was connected across the voltage under observation, so that the actual voltage measured by the curve tracer was a fraction k of the voltage at the same time, where k was proportional to I. Thus, overall, the actual recorded value was proportional to VI, the instantaneous power. To quote James G Biddle's catalogue (circa 1899), 'Thus are the power-curves accurately drawn to a pre-determined scale, and the labor of multiplying the corresponding ordinates of current and electromotive force is avoided'. One wonders!

This apparatus was obviously rather bulky, and Professor Rosa also developed the more portable version shown in figure 3.29 [72–74]. Here the knob at the right was used to move the slider over a helically wound potentiometer wire, and the mechanical action of pressing the stylus knob made a mark on the drum through a typewriter ribbon and also latched the drum and the contact seen on the left onto the next position. The contact-making mechanism of both models, of course, required to be connected by a shaft to the alternator producing the voltage under observation. By this time (about 1905) the Willyoung Co. had been bought out by Morris E Leeds who changed its name, first to the Morris E Leeds Co., and eventually in 1903 to the Leeds and Northrup Co. whose name can be seen on the base plate of the instrument. According to F A Laws' book [72] (first edition 1917: second edition 1938), 'The Rosa curve tracer furnishes the most accurate apparatus yet devised for mapping periodic electrical phenomena'.

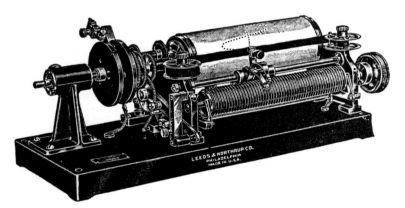

Figure 3.29   The portable version of Rosa's apparatus. The potentiometer is wound helically around the cylinder at the front [72]. Courtesy: McGraw-Hill Book Co.

A somewhat similar manual tracing arrangement was used by the German Engineer, Dr Rudolph Franke, then at the Technische Hochschule, Hannover, in 1899 [75–77]. This apparatus is shown in figures 3.30 to 3.32. The rotating commutator which was driven by the alternator under test, or by a synchronous machine, was furnished with five segments, one of them occupying 120° of arc, the others 60° each. The only active segment was the large one which momentarily connected brushes B_1 and B_2 once in every rotation. The only function of the other segments, which were too small to allow contact between B_1 and B_2 was to ensure that the brush contacts rubbed on the same material throughout the rotation, as opposed to the more usual practice of setting a metal contact into an insulating disc. This

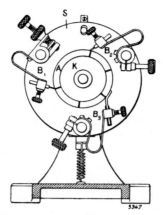

Figure 3.30   Commutator type of contactor used by Franke [76].

constancy of the friction was said to depress any tendency of the brushes to vibrate or judder, and hence gave more reliable operation. The third brush holder B_3 carried a small pad soaked with petrol or paraffin to keep the metal surfaces free from grit and dust. The contact was set to different angular positions by rotating the plate on which the brushes were mounted.

The contact mechanism was connected in series with a mirror galvanometer, and the trace was recorded using the system shown schematically in figure 3.31. The angular position of the recording drum was determined by linking it to the contact plate with the cord S, and the galvanometer light spot was arranged to deflect along the axis of the drum. The complete apparatus can be seen in figure 3.32. With the right hand, the operator set the position of the contact using the knob at the rear acting

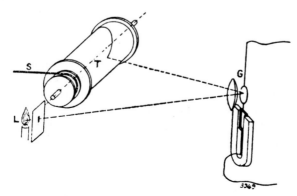

Figure 3.31   Franke's recording system, showing linking of recording drum to contact mechanism by cord and pulley [76].

Figure 3.32 Franke's complete instrument, showing method of operation [75]. Photograph: Science Museum (396/86).

through a worm screw. At the same time, the left hand was engaged in turning another knob which caused a pencil to move along the axis of the drum. The operator kept his eye on the light spot, and ensured that the pencil always followed its movements. Some of the curves traced by this apparatus are reproduced in the papers cited above, and these exhibit a degree of 'shakiness', as might be expected from a piece of apparatus of this sort. It must have demanded a high degree of concentration on the part of the observer, not to mention good hand-to-eye co-ordination. It is perhaps worth mentioning here that the Cambridge Instrument Company produced an instrument called a 'Brearly Curve Tracer' around 1910 [78,79]. This was designed by Harry Brearly of Sheffield, a Metallurgist, to be used with a mirror galvanometer system, and it allowed one to transfer the position of a light spot directly onto a recording drum by following its movements manually. It was intended primarily for use in recording recalescence (heating and cooling) curves for steel, but as the maker's catalogue says 'Its usefulness is extended to the recording of curves in any experiment in which the deflections of a galvanometer vary with time'. No doubt it was pressed into service from time to time in making permanent oscillographic records. Manual tracing of a light spot with a pencil had also been suggested in 1895 by Ayrton and Mather of the City and Guilds College, London [80] who had been experimenting with a synchronously driven contact system.

Marjan Lutoslawski [81], in 1896, had constructed a system having a rotating drum for recording the trace—a system, in fact, very similar to that of Franke which followed a few years later. However Lutoslawski's galvanometer was equipped with a long pointer carrying a stylus which made a mark on the drum, thereby eliminating the need for manual tracing of the spot. In 1898, Professor H L Callendar of University College London, and later of Imperial College, adapted his recently invented chart recorder [82,83] for the fully automated tracing of electrical waveforms. The chart recorder was a simple self-balancing potentiometer linked to a pen which made a trace on a moving band of paper. In his waveform tracer this was used in conjunction with a worm-driven contact mechanism so that the curve was produced automatically without the need for constant attendance. The disadvantage of his system was that it took no less than one hour to draw out a complete cycle of the waveform. It may well have allowed one to go away for a cup of tea while the operation was being carried out, but it certainly did not meet the criticism mentioned earlier about the constancy, or otherwise, of the waveform during the time of measurement. As an unrelated point of passing interest, Professor Hugh Callendar was also the author of 'The Manual of Cursive Shorthand', and 'A System of Phonetic Spelling Adapted to English', both published in 1889.

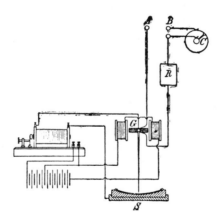

Figure 3.33 Automatic recording of contact voltage using spark photography [85].

A much speedier automatic system was that introduced by Mr Frederick Drexler of the Oerlikon Company in Vienna [84,85]. This apparatus, which dates from 1896, is shown in figure 3.33. The moving contact C was mounted on the shaft of an electric motor, which could be either AC or DC driven. The alternating waveform whose shape was to be traced was applied to the coil of a galvanometer via the contact. The magnetic field of the galvanometer was produced by current from a subsidiary battery as shown.

Imagine for the moment that the time for one revolution of the contact motor was precisely the same as one period of the alternating waveform. This would mean that the waveform would be sampled at the same instant in each cycle, and the galvanometer would give a constant reading. If the contact-motor speed were now allowed to drift very slightly, the sampling instant too would drift through the waveform, and the galvanometer pointer would oscillate at the 'beat' frequency, its tip tracing out the AC waveform. The method of recording its movement without introducing any frictional errors is quite interesting. The needle was connected to one terminal of a high-voltage induction coil, and the other terminal of the coil was connected to a trough-shaped metal plate situated near the end of the needle. A piece of photographic film was clipped into the trough and the silent electric discharge passing between the needle and the plate decomposed the silver salts and caused an image to appear when the film was developed. The plate was caused to move forward (out of the paper in figure 3.33) so that a voltage/time graph was drawn. The speed of revolution of the motor was selected so that the whole tracing process was completed in under two seconds. The idea was a very good one, but if the published results are typical, the electrical discharge method of recording produced a very fuzzy and ill defined trace. Furthermore, as was pointed out in the course of a discussion at the Institution of Electrical Engineers in 1903 [86], 'The slip of an asynchronous motor is too variable and too great to be employed satisfactorily for direct stroboscopic recording, and (Drexler's) arrangement does not allow of the exact determination of the phase relations of curves traced successively, since the slip is not systematically connected with the rotation of the recording cylinder'.

The same principle of 'phase drift' was employed with greater precision in the apparatus built by Wilhelm Peukert in 1889 [87,88]. At the heart of the system was a train of gear wheels shown as R_1 to R_4 in figure 3.34. The wheels R_2 and R_3 were mounted rigidly together on the same shaft. Wheel R_1 has n teeth, R_2 $n + 1$, R_3 has n, and R_4 has $n - 1$. With these ratios, the speeds of the first and last shafts will be given by

$$\frac{S_F}{S_L} = \frac{n}{n + 1} \frac{n}{n - 1} = \frac{n^2}{n^2 - 1}$$

a ratio not far from unity.

Thus a contact mechanism mounted on the last shaft as shown will slowly 'creep' in phase relative to the first shaft. If that first shaft were the shaft of an alternator, then a galvanometer connected through the contact to its output voltage would slowly move through the voltage cycle. In fact Peukert's apparatus was supplied with two sets of contacts, and two galvanometers were mounted one above the other using a common scale. The phase difference between two waveforms could then be appreciated by the relative movements of the two pointers along the scale.

One of the most interesting and successful of the fully automatic contact methods was the 'ondograph' invented by Monsieur E Hospitalier, and demonstrated by him at the Institution of Electrical Engineers in London in 1903 [89–95]. This instrument was constructed by the Compagnie pour la Fabrication des Compteurs et Matériel d'usines à Gaz, Paris, and was available commercially. A version of it was made by H Tinsley and Co. in Britain.

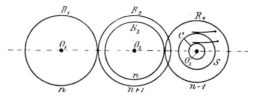

Figure 3.34 Peukert's system of gear wheels to produce progressive advance of the contact phase. The speed ratio of the first and last shafts is $n^2 : (n^2 - 1)$ [87].

Edouard Hospitalier, who was Professor at the École de Physique et de Chimie Industrielles in Paris, was the author of several electrical textbooks and was also the editor of the journal *L'Électricien*. He was elected a foreign Member of the Society of Telegraph Engineers and of Electricians (later the Institution of Electrical Engineers), London, in 1881, having been proposed by the famous W H Preece, and seconded by J Aylmer and W E Ayrton.

Incidentally, for the benefit of those purists who objected to the mixed French/Greek etymology of the word 'ondograph', he did also permit the use of the alternative terms 'cymatograph' or 'kymatograph'. The special rotating contact incorporated in this instrument is shown in figure 3.35. The shaded area represents insulating material; the blank, conducting material. As the shaft rotates, the brushes D and B are connected together for a brief

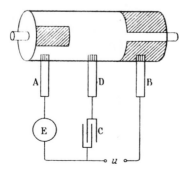

Figure 3.35 The contact maker of Hospitalier's ondograph [89].

instant so that the capacitor is allowed to charge up to the applied voltage *u*. For the rest of the time, the contacts A and D are connected so that the capacitor is discharging through the moving-coil meter E. The complete instrument is shown in figure 3.36. When starting it up, the synchronous motor on the left was brought up to correct speed by means of a handle fixed to the shaft seen at the centre of the picture. A stroboscopic disc which is also visible was provided to help in this. The alternating supply was then connected so that the motor continued to rotate at the correct speed. The motor was used to drive two gear trains, one rotating the contactor shaft, the other turning a drum upon which was fixed a piece of paper. The pointer of the meter rested upon the drum and drew an ink line on it. The drum gearing was such that it turned once in every 3,000 cycles of the incoming supply. If the contact also rotated at this speed, then the samples would be taken at the same point in each cycle and constant deflection would result. However, the gearing was such that the contactor rotated only 3×999 times during this interval, so that there was a slow drift between the drum and contactor speeds. The waveform was thus sampled at successively advancing instants and was drawn out on the drum. In fact, with the gearing ratios quoted, three complete cycles were traced.

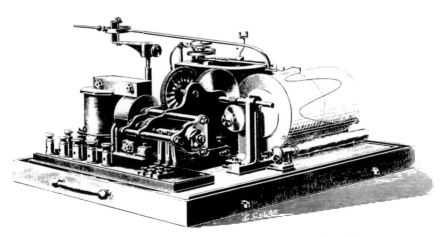

Figure 3.36   The ondograph (or kymatograph) [89].

The ondograph was also arranged so that it could draw instantaneous power curves. For this type of measurement, a dynamometer meter was substituted for the permanent magnet type. Its fixed coils carried the current and the moving coil was operated with the capacitor in the normal way to measure voltage. When used in this way, the approved terms for the instrument were the 'puissancegraph' or 'electrodymograph'.

The method of driving the ink pen is rather interesting. In order to inscribe a straight line along the recording cylindrical drum an infinitely long pointer would be required. It is very difficult to construct a moving-coil movement capable of driving a very long arm directly because of inertia and pen friction, and so a compromise was devised. The pen itself was carried on the end of a long lightweight arm, pivoted and delicately balanced on a gimbal mechanism—or a 'Cardan' joint as it was referred to in contemporary accounts. The end of the actual meter pointer carried a pin which moved in a slot cut in the pen arm resulting in almost linear movement of the pen along the recording drum. Because of the slight residual curvature the instrument was supplied with special curvilinear graph paper to wrap around the drum. This can be seen in figure 3.37 which shows a typical voltage curve obtained by use of the ondograph.

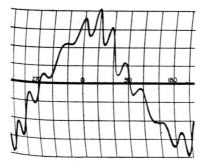

Figure 3.37 A typical waveform record produced by the ondograph. Note the special curvilinear recording paper [89].

There were, as we have seen, several commercial versions of the Joubert contact system on the market, Hospitalier's ondograph being one such. However, most of the engineers whose names have been mentioned built their own contact systems in their own establishments and for their own use, without any immediate thoughts of producing a commercial instrument for sale as such. They were often constructed in the first instance with a particular application or investigation in mind. Thus, to take a few examples, the contact apparatus constructed by B F Thomas was intended specifically to aid in the investigation of the behaviour of an arc lamp connected to a Brush generator; that of H J Ryan was to assist in transformer design; that of L Searing and L V Hoffmann to look into the effects of magnetic saturation upon the waveform produced by a generator; and F Drexler's version was to be used in monitoring the performance of the public lighting system in Vienna. When the contact mechanisms *had* been built however, their owners then had in their possession general purpose tools of great utility which could be employed in all sorts of other investigations involving alternating currents, much as a cathode ray oscilloscope today can be used for a multitude of different purposes. Other experimenters such as J A Fleming seem, as far as can be inferred from their published papers, to have set out with the idea of building a general purpose instrument, intending to use it for a number of different projects which they had in mind.

The contact method was capable of being used with measuring systems of considerable elaboration and sophistication. Again, simply as one example, we may quote the work of H Rupp who, in 1900, published details of some very subtle arrangements of rotating machines and contacts designed to study the EMF and current waveforms in the rotors of induction motors with different degrees of slip in the machines [96].

3.4   Rotating contacts used for other purposes

Once rotating contact makers had been introduced into laboratories and mounted on the machine shafts, experimenters began to find other uses for them besides the plotting of waveforms. A Wilke, for example, used two contact discs to ensure that the shafts of two machines were running in synchronism, and to measure their relative phase relationship [97]. The contact discs were mounted on the two shafts as shown in figure 3.38 and a battery and telephone receiver were connected in series with them. Only

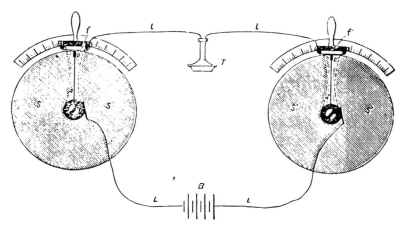

Figure 3.38   Two contact discs used to measure the phase relationship between two rotating shafts [97].

when the swinging contact brushes were set to an angular difference equal to the phase displacement of the shafts would clicks be heard in the ear piece.

Major P Cardew, who is perhaps best remembered for the hot-wire voltmeter which he developed in 1884, made use of a double set of contact brushes with one single rotating disc in order to measure the phase angle between the voltage and the current in a circuit [98,99]. One swinging brush was arranged so as to measure the voltage; this was first set so that its associated galvanometer read zero, thus indicating the position of the zero crossing of the voltage waveform. The second brush was likewise set to measure the zero crossing in the current waveform. The difference in angular settings between the two brushes thus gave a measure of the separation between the zero points of the waveforms—i.e. measured the phase difference.

Sometimes the contact systems were modified in various ways in order to perform particular specialised types of measurements. It will be recalled that when describing the curves obtained by Fleming (figure 3.17) it was mentioned that since the voltage produced in an alternator is proportional to the rate of change of flux $d\varphi/dt$, the actual flux distribution curve may be obtained by integration of the voltage waveform. Although perfectly straightforward in theory, this was actually a very tedious computation to carry out, involving as it does the measurement of successive areas under the curve. In 1900, F Townsend suggested a rather interesting method of measuring the flux curve directly [100,101]. He used a contact disc in which the contact closure was not merely momentary as in the usual system, but instead occupied $180°$ of the rotation. Under these conditions, assuming that the contact is made for an interval t_1 to $t_1 + T/2$, where T is the period of the waveform, and also the time for one revolution of the disc, the indicating instrument will settle to give a deflection θ proportional to the average value of the waveform which it receives, i.e.

$$\theta \propto \frac{1}{T}\int_{t_1}^{t_1+T/2} E(t)dt$$

and since the voltage waveform $E(t) \propto d\varphi/dt$, then

$$\theta \propto \frac{1}{T}\left[\varphi(t)\right]_{t_1}^{t_1+T/2} \propto \frac{1}{T}(\varphi(t_1 + T/2) - \varphi(t_1)).$$

If it is assumed that the flux distribution in the machine is such as to produce an alternating wave of symmetrical form (i.e. each half-cycle is the mirror image about the time axis of its predecessor) then

$$\varphi(t_1 + T/2) = -\varphi(t_1)$$

and the deflection is simply proportional to $\varphi(t_1)$. Thus by varying the contact position (varying t_1), the flux waveform can be measured directly.

Yet another use of the contact maker was for the analysis of the alternating waveform into its Fourier harmonic components [102]. Assume that the waveform can be represented by the Fourier series

$$f(t) = C_1 \sin(\omega t + \alpha_1) + C_3 \sin(3\omega t + \alpha_3) + C_5 \sin(5\omega t + \alpha_5) \dots$$

the fundamental period of the waveform being $T = 2\pi/\omega$. (It was generally assumed that the waveforms of interest contained no even harmonics, and this assumption will also be made here. It should be noted, however, that the theory which will be developed is also perfectly valid for even harmonics if required.). For the sake of illustration, let the normal type of contact maker be replaced by one which makes contact at three equally spaced times per cycle of the fundamental. The waveform samples obtained will then be:

$$S_a = C_1 \sin(\omega t_1 + \alpha_1) + C_3 \sin(3\omega t_1 + \alpha_3) + C_5 \sin(5\omega t_1 + \alpha_5) \dots$$

$$S_b = C_1 \sin[\omega(t_1 + T/3) + \alpha_1] + C_3 \sin[3\omega(t_1 + T/3) + \alpha_3]$$
$$+ C_5 \sin[5\omega(t_1 + T/3) + \alpha_5] \dots$$

$$S_c = C_1 \sin[\omega(t_1 + 2T/3) + \alpha_1] + C_3 \sin[3\omega(t_1 + 2T/3) + \alpha_3]$$
$$+ C_5 \sin[5\omega(t_1 + 2T/3) + \alpha_5] \dots .$$

Expressing these in terms of the phases of each individual harmonic (and remembering that a given quantity of *time* represents a different *phase*shift for each harmonic)

$$S_a = C_1 \sin(\omega t_1 + \alpha_1) + C_3 \sin(3\omega t_1 + \alpha_3) + C_5 \sin(5\omega t_1 + \alpha_5) \dots$$

$$S_b = C_1 \sin(\omega t_1 + \alpha_1 + 120°) + C_3 \sin(3\omega t_1 + \alpha_3 + 360°)$$
$$+ C_5 \sin(5\omega t_1 + \alpha_5 + 600°) \dots$$

$$S_c = C_1 \sin(\omega t_1 + \alpha_1 + 240°) + C_3 \sin(3\omega t_1 + \alpha_3 + 720°)$$
$$+ C_5 \sin(5\omega t_1 + \alpha_5 + 1200°) \dots .$$

(For ease of understanding, the phases in these equations have been left in degrees although strictly speaking, for dimensional correctness, they should be in radians.)

It will be recalled that three values of a sinusoid taken at $120°$ intervals add up to zero. The galvanometer connected to the contact maker will settle at a deflection which is the average of the samples it receives. In this case, the average of the samples of the fundamental component will be zero, as will those of the fifth, seventh, ..., etc. On the other hand the average of the third harmonic will be $3C_3 \sin(3\omega t_1 + \alpha_3)$. The principle is thus established. If a contact disc is supplied with three equally spaced contacts around its rim, then only the third harmonic will contribute to the deflection of the galvanometer. If the contact brush position is advanced in steps in the usual way, it is possible to plot the third harmonic alone so that its amplitude and phase can be determined. A contact maker with five contacts can be used for the fifth harmonic, seven for the seventh and so on.

A word of warning; with three contacts, the sixth, ninth, ... etc harmonics will cause deflection as well as the third. Similarly, the five-contact disc will respond to the fifth, tenth, fifteenth, ..., harmonics. With most commonly found waveforms, the amplitudes of the harmonics diminish rapidly with increasing frequency and it is usually reasonable to assume that with the alternating waveforms normally encountered, only the third, fifth and seventh harmonics are present to any appreciable extent, and the method is usable and provides results of adequate accuracy.

As a variant of this, the contact disc can be provided with a series of on/off segments in the manner of Townsend's integrating system mentioned above, and provided that the waveform is differentiated before measurement, a set of discs having different numbers of segments can again provide a harmonic analysis of the waveform [103]. An elaborate contact drum or commutator working on this principle was constructed by T R Lyle [104] in 1903—see figure 3.39. This was able to plot the third, fifth and seventh harmonics. It was furnished with a pulley, string and photographic-plate arrangement of the usual sort to produce a permanent record of the harmonic waveforms, and a typical set of curves obtained by this system is shown in figure 3.40. This shows the actual waveform under analysis, together with separate plots of the third harmonic (magnified by a factor

Figure 3.39 Lyle's commutator system for harmonic analysis [104].

of three times), the fifth (magnified five times) and the seventh (magnified seven times).

Interest in this particular type of analysis seems to have been rekindled, for some reason, in the 1930s, and it will be noticed that the papers of Dannatt and of Gall [102, 103] date from this period. At this time also the contact disc system was used in conjunction with a valve oscillator to measure harmonic content of a waveform [105], the contactor's function being to lock the phase of the oscillator.

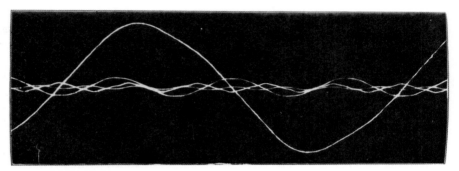

Figure 3.40 Typical set of harmonic traces produced by Lyle's commutator analyser [104].

3.5 The observation of repeated transient waveforms

Of course, as previously mentioned, all these contact methods, whether for waveform determination or for harmonic analysis, are only suitable for repetitive, periodic waveforms. They cannot be applied to 'one off' transient waveforms. On the other hand, if the transient waveform can be arranged so as to be repeated continually, then it becomes in effect a periodic waveform, and the contact method can be applied. The experiments carried out by J C Hubbard of Clark College, New York [106] in 1913 to determine the frequency of oscillation of a tuned circuit represent a very simple example of such an approach. The circuit which he used is illustrated in figure 3.41. *L* represents an inductor and *K* a capacitor. In the initial state, the parallel tuned circuit was connected in series with a battery *B* and a resistor *R*. A falling weight was arranged to strike the horizontal part of lever 1, thereby opening the battery circuit and initiating a transient oscillation in the tuned circuit. The weight then struck lever 2, and this pivoted about its centre, eventually striking lever 3, and putting the electrometer instrument in connection with the circuit. (The switch a had been used to reduce the electrometer charge to zero before starting the measurement, and at this stage would have been open.) This connection was only momentary, since the lifting of lever 3 disconnected the electrometer once more. The result of all this was that the electrometer was briefly in connection with the circuit, and a short sample of the capacitor voltage was stored on its plates. Its deflection indicated the magnitude of this voltage. By varying the vertical distance between lever 1 and lever 2, the voltage sample could be taken at varying intervals of time after the initiation of the transient and, if required, a complete picture of the waveform could be built up by repetition of the experiment. In fact, Hubbard was interested primarily in the frequency at which the circuit oscillated, and he concentrated on the determination of the zero crossings in the waveform—i.e. those instants at which the electrometer reading was zero. Measuring apparatus of this sort was often referred to as the 'drop chronograph'.

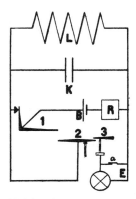

Figure 3.41 Hubbard's drop chronograph [106].

Alfred Hay of Nottingham (later to move to the Indian Institute of Science, Bangalore) in 1894 used a pair of rotating contact makers to observe and study 'impulsive current rushes in inductive circuits'—in other words, the transient phenomena which take place when inductors or transformers are suddenly connected to the AC mains [107]. Both contact discs were mounted on the shaft of the alternator so as to rotate synchronously. One of the discs, seen in figure 3.42, was designed to create the transient repeatedly. It was constructed of wood, having a metal rim divided into two segments by means of the insulating distance pieces DP. The larger segment was the active one; the other was provided simply for the sake of 'true running' as the inventor expressed it. The larger segment was in connection with a slip ring with which contact was made by the brush B_2. The curved bar GS (the letters standing for 'graduated scale', the bar being so provided) carried two adjustable contacts B and B_1. The circuit arrangement was as shown, so that the mains waveform was applied to the inductor or transformer as long as the larger segment made contact with either brush.

The position of B controlled the instant of making contact, the position of B₁ the instant of breaking contact. Thus a carefully controlled segment of the mains waveform could be applied to the circuit. The second contact maker disc was of the usual Joubert type with a swinging brush, and was used to sample and delineate the various current and voltage waveforms in the normal point-to-point manner. Interestingly enough, a very similar system was used in 1891 by Grawinkel and Strecker to determine the response of a transmission line to a repeated Morse dot [108,109].

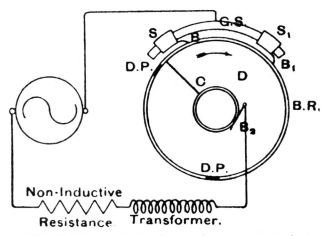

Figure 3.42 Hay's method of creating a repeated transient phenomenon, thus enabling its observation by the Joubert method. A second contact disc, mounted on the same shaft as that above was used to take the samples of the waveform [107].

There was one other major field of research where the sampling of repeated transient waveforms assumed some importance. During the second half of the nineteenth century, there was considerable interest in the electrical phenomena which accompany the actions of nerves and muscles in the body, and several eminent physiologists constructed apparatus to observe the electrical signals generated when muscles were stimulated in various ways. These instruments were referred to as 'rheotomes'.

Words derived from the Greek root RHEO-, meaning a stream or current, have a long and honourable lineage in electrical science. In the Bakerian Lecture he presented to the Royal Society in 1838, Charles Wheatstone pointed out that the word RHEOPHORE (from the Greek word 'to carry') had been used by Ampère 'to designate the connecting wire of a voltaic apparatus' [110]. He also mentioned that the word RHEO-METER (from the Greek verb 'to measure') had been adopted by French writers to describe the instrument which later became known as the galvanometer. Wheatstone himself proposed a list of new RHEO- words. Some of these never gained currency, and the rest have fallen into disuse—with the single exception of RHEOSTAT (a variable resistor used to set the value of a current). The word RHEOTOME is found only very occasionally in the electrical literature of the time, but was quite widely used in medical circles. It was derived from the verb 'to cut' and was defined originally as 'an apparatus which periodically interrupts a current'. This definition seems to have been extended gradually to embrace the function of sampling and measurement.

In 1886, Emil DuBois-Reymond, Professor of Physiology at Berlin, published a paper entitled 'On the time required for transmission of volition and sensation through the nerves' [111,112]. As its title indicates, this paper was concerned with the velocity of propagation of electrical impulses through the nerves of the body. It is interesting to note, in passing, his reference to an earlier speculation made in 1762 which suggested that 'the velocity of the nervous agent ought to bear the same proportion to that of the blood in the aorta as the width of the aorta to that of the nervous tube' In other words, the thinner the conducting channel, the faster the movement through it. This speculation led to the result that the velocity of propagation of nervous pulses must be 120,000,000 miles s^{-1}, or over 600 times the velocity of light!

DuBois-Reymond's apparatus was a modification of a method used by Helmholtz, which was in turn based on a device known as Pouillet's chronoscope which had been used to determine the muzzle velocity of a bullet. The principle was that

> a short pulse of current through a galvanometer coil gives it a single impulse, and thereafter it slowly settles back to its zero reading. If the current is constant, the initial velocity will be proportional to its duration, and that duration can be inferred from the deviation of the needle. The galvanometer is therefore transformed into a chronoscope (a time measuring device).

In modern parlance, the galvanometer was used ballistically; its throw was a measure of the total charge flowing, and if the current were constant it would effectively measure the time for which that current was applied. In Pouillet's bullet measurement the action of the gun-cock striking the cap was made to complete the circuit, and the emergence of the bullet from the end of the muzzle was made to break it again so that the time of travel down the gun, and hence the velocity, could be determined.

Figure 3.43 shows the apparatus constructed by DuBois-Reymond. A frog's muscle g and its associated nerve fibre AB are suspended from a clamp c. The lower end of the muscle is fixed to a lever pivoted at k and it is kept in a suitable state of tension by the weight placed on the suspended balance pan. The coils pc and sc are the primary and secondary windings of an induction coil. The primary current is supplied by the battery B_2 via the contact q which is normally closed. Interruption of the primary current causes a pulse of voltage to appear in the secondary coil and this is applied to the nerve at A or B, or across the muscle itself according to the setting of the two-pole three-way switch.

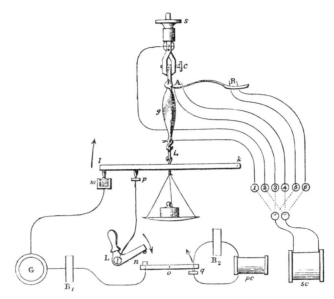

Figure 3.43   The DuBois-Reymond rheotome [111]. Courtesy: Royal Institution.

In operation, the handle L was moved over sharply, thereby opening q and producing the impulse in sc, and at the same time closing the circuit of battery B_1 through the galvanometer G, the contact dipping into the mercury cup m, and the point-and-plate contact p. The stimulation of the nerve caused the muscle to contract, which lifted the lever and broke the galvanometer circuit. The throw of the galvanometer was then a measure of the time between activation of the circuit and reaction of the muscle. After stimulation, a muscle returns to its original length, and this would cause the lever to fall, so completing the galvanometer circuit again and producing further deflection. This was avoided in the following way: before the measurement was made, the mercury cup was lowered very carefully so that capillary action drew up a cone of mercury from the surface to the point. When the lever rose, the mercury fell back and resumed its normal meniscus shape so that when the lever dropped back the point was unable to reach the mercury and the circuit remained uncompleted. To operate the circuit a second time the cup had to be raised and lowered to re-establish the capillary cone.

The velocity of propagation of the nerve impulse along the nerve fibre could be determined by performing the experiment with the impulse first applied at A, and then repeating it with the impulse applied at B. The difference in response time was a measure of the time taken for propagation of the impulse from B to A.

As described here the apparatus was simply a timing arrangement, but the author went on to describe further experiments which enabled him to produce a graph showing displacement of muscle versus time. These involved attaching a light stylus to the end of the lever and allowing it to make a trace on a moving smoked glass plate or a rotating drum, an arrangement which was referred to as a 'myographion'.

Another piece of apparatus, details of which were published in 1878 [113], was the so-called 'fall-rheotome' used by Professor L Hermann, a device which was similar in principle to Hubbard's 'drop chronograph' of figure 3.41. When the surface of a muscle is damaged, a potential difference is produced across it and Hermann's apparatus (figure 3.44) was designed to investigate the magnitude of this voltage and the delay in its appearance. Here the muscle under investigation is stretched across an ebonite block Q. A weight is allowed to fall between slides from a height of about four feet, the path of its descent being indicated in the figure by the dotted lines. One side of the weight is covered with shagreen (the rough belly skin of a shark or ray) so that as it falls it strips off the surface layer of the muscle. A galvanometer is connected across the muscle via the switches g and g'. As the weight falls it first closes the circuit by striking the lever of g and moving the contact to the upper segment. A moment later it strikes switch g' and the circuit is opened again.

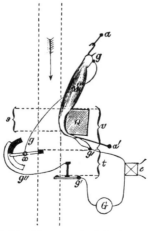

Figure 3.44 Hermann's fall-rheotome [113].

The positions of the block Q and the pivot of switch g could be moved up and down to vary the instant and duration of closure of the galvanometer circuit. These times could be determined from the equations of motion of the falling weight. As an indication of the sorts of times involved, the duration of closure was said to be 3.4 ms.

The apparatus is capable of taking a sample of known duration at a given time after the phenomenon under observation has been initiated. In principle, the waveform of the muscle voltage could be determined by repetition of the measurement for different time delays, but as far as the present author can determine there is no evidence to prove that Hermann actually used it in this way. The main thrust of his experiment seems to have been to determine the time taken for the voltage to develop after injury to the muscle. The conclusion was that after the weight had fallen a certain potential difference would be observed, but if the weight were allowed to fall a second time (without further injury) a greater value would be obtained thereby demonstrating the need to wait a while before the full effect of the injury became apparent. The apparatus was also modified to determine the effect of an electrical stimulation of the muscle nerve by removing the block Q and substituting another switch operated by the falling weight. This opened the primary circuit of an induction coil as in the Reymond apparatus. Hermann observed the output voltage across the muscle at various time delays. He found that there was no response for about 2 ms, and after that there was a response of one polarity followed by reversal after 8 ms or so.

It will be recalled that Charles Wheatstone's original definition of the word RHEOTOME was 'an apparatus which PERIODICALLY interrupts a current'. This description does not really fit the two pieces of apparatus just described except inasmuch as the experiment could be repeated by operating the lever or allowing the weight to fall over and over again. In the account of Hermann's work previously quoted there is reference to the fact that he also constructed a rheotome which could apply pulses periodically every sixtieth of a second. In fact, Professor Burdon-Sanderson (of whom more later) made a clear distinction between the 'fall rheotome' of Hermann and a 'differential or repeating rheotome' which was constructed by Julius Bernstein of the University of Halle, near Leipzig, which was described in Bernstein's book published at Heidelberg in 1871 [114]. Figure 3.45(*a*) is an elevation showing the construction of this device and figure 3.45(*b*) is a plan view which also shows the electrical connections. It had a central vertical shaft driven via a pulley by an electric motor. Two horizontal arms were attached to the shaft, arranged diametrically opposite to one another. One of them carried a contact p which projected below the arm and struck the stretched wire d as it passed in each revolution. This contact allowed a short pulse of current to flow from the cell K into the primary winding P_1 of an induction coil. When the shorting switch S_2 was opened the secondary pulse was applied to the nerve fibre NN through the electrodes rr. The electrical pulse transmitted through the nerve was picked up by the contacts N and T. Two contact points p_1 and p_2, connected together and mounted at the end of the other rotating arm, dipped into the cups of mercury q_1 and q_2 and took a sample of the nerve voltage once in each revolution as long as the shorting switch S_1 was open. This voltage was measured by means of the slide wire potentiometer arrangement shown at the top right hand of the diagram. D was a Daniell cell, Q a simple mercury on/off switch and RR the potentiometer wire. W was a reversing switch to allow measurement of voltages of both polarities. Any steady DC potential which happened to exist across NT could be measured by insertion of the link b between points ss. Interestingly enough, Bernstein referred to this potentiometer system as the 'rheocord', and its action was explained in terms of compensating currents.

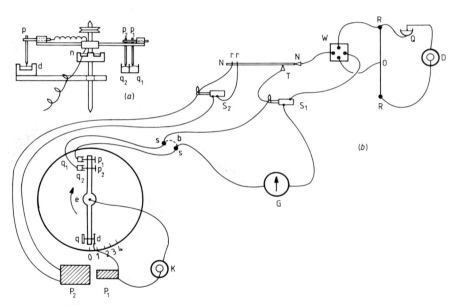

Figure 3.45 Bernstein's rotary (or repeating) rheotome [114].

Thus the phenomenon was triggered and sampled once in every rotation. The instant at which the impulse was produced could be varied by altering the position of the wire d, which was mounted on a swinging arm, and setting it at various points around the graduated circle. The complete waveform could then be delineated. The diagrams in Bernstein's book (e.g. figure 3.46) show that this was a carefully engineered instrument, and he spoke of measured times such as 0.0005797 and 0.0006304 s. Such figures would seem to imply an extraordinary degree of accuracy and great faith in the precision of his experimental apparatus. Let us assume for the moment that it would be possible to set the contact at intervals of one degree around the circle. If it is assumed that this spacing represents a time interval of 0.0005 s, then, this would correspond to a rotational speed of 5.55 revolutions per second or 333 rpm, which is by no means excessive. The present author has been unable to find in Berstein's book any mention of the speed of rotation actually used, but an increase of speed to 3,000 rpm together with careful vernier setting of the contact position could have made measurements in time differences to within 10 µs a reasonable proposition. On the other hand, a muscle is unlikely to be able to respond separately even to five impulses per second, so it is safe to conclude that the figures quoted are of quite spurious accuracy as far as muscle measurements are concerned.

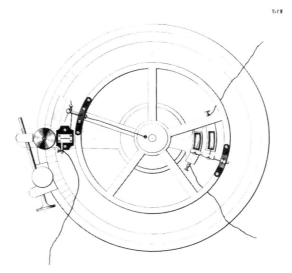

Figure 3.46   Plan view of Bernstein's rheotome [114]. Photograph: British Library (7407ee.14,Taf.II).

This type of rotary rheotome was further developed by Professor Sir John Scott Burdon-Sanderson, a physiologist of renown whose early reputation had been established by his studies of the spread of infectious diseases, and who turned in the 1880s to investigations of electrical phenomena in nerves and muscles, and also in the well known plant Venus fly trap [115,116]. His rheotome is shown in the rather beautiful engraving of figure 3.47. A circular iron base plate is supported on three levelling screws and bears two pillars which support a horizontal bar. A point mounted at the centre of the plate and a further one projecting downwards from the bar act as bearings for a vertical shaft 5 cm long. The shaft is surrounded by a circular disc of vulcanite in which are cut eight wedge-shaped depressions to contain pools of mercury. Alternate pools are connected together by platinum wires thus forming two sets of interleaved mercury segments, each of which rises above the vulcanite surface, forming a meniscus due to the surface tension. These two sets of pools are connected to the terminals whose wires are labelled G and G', seen at the rear of the apparatus. The vulcanite is further surrounded by an annular ring, also of vulcanite, in which are cut eight square mercury pools. Each pool can be put in contact with a circular mercury trough which is also on the ring by removal of a little vulcanite stop piece. Only one of these pools is shown filled with mercury in the diagram. Contact is made with the trough by means of a wire point h mounted on the insulated terminal B. The whole annular ring can be rotated relative to the disc by means of the tangent screw and graduated knob seen at the front of the apparatus. The vertical shaft bears three horizontal arms insulated from the shaft by a non-conducting bush. The longest arm, seen on the left, carries an amalgamated gold point which dips into the outer mercury pools. This arm has an insulating segment but is connected to the axle, which is in turn connected to the circular iron trough containing mercury mounted at the top of the axle, and thence via the bent point to the horizontal bar and the terminal B' on the iron plate (at the front). The result is that as the shaft is rotated

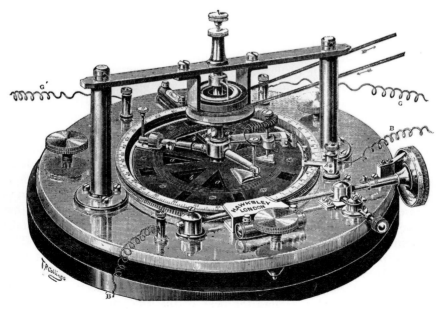

Figure 3.47 Burdon-Sanderson's rheotome [115]. Courtesy: Royal Society.

by means of the pulley, the terminals B and B' are brought into electrical contact once in each revolution as the point touches the mercury surface.

The two short arms are 45° apart and are in metallic contact with each other, but insulated from the shaft itself and from the other point. One (e) has a point which dips into the mercury at the outer edge of the wedges; the other (f) has a point which can be moved radially so as to cut the mercury wedges at a smaller distance from the centre. When these two points are in adjacent pools the terminals G and G' are in contact.

In use, the pulse of exciting current is applied through the terminals B and B', the instant of application being determined by the setting of the tangent screw and annular ring. The galvanometer circuit is connected to G and G'. The duration of the galvanometer connection is governed by the radial position of the point f. It thus provides a method of applying a trigger pulse and of taking a sample of variable duration at a defined instant of time. According to the requirements of the particular experiment one outer and two inner pools can be filled with mercury as shown in the diagram, or for repetition at a faster rate all the pools can be used. Unfortunately, Burdon-Sanderson in his account of this rheotome gives no details of the speed of rotation he used or of the time intervals observed. However this was clearly a well thought out and carefully constructed instrument and the remarks made previously about the accuracy of Bernstein's rheotome would also apply here.

Thus, to sum up this chapter, the point-to-point or sampling method, whether invented originally in Europe by Lenz, Wheatstone or Joubert, or in the United States by Thomas, appeared in many different forms, and remained the principal method of waveform delineation for fifteen or twenty years until superseded by the oscillographs at the turn of the century. The oscillographs in their turn yielded supremacy to the cathode ray oscilloscope, but the cathode ray tube itself was pressed into service as an inertialess contact maker by Ernst Lübke of Göttingen in 1919 [117]. The deflection coils of the tube were connected to two sinusoids having a 90° phase relationship so that a circular Lissajous figure was produced at the screen end of the tube. In the words of the author 'the cathode stream was rotated in the form of a hollow cone'. Instead of the usual fluorescent screen, the rays fell on a mask in which was cut a radial slit so that a short pulse of rays was allowed through once in each cycle. This pulse of rays entered the space between two electrodes and ionised the gas there, so that the conductivity increased markedly. This was then used as a contact to take a sample of the waveform whose shape it was desired to observe. This waveform and the sinusoids which were producing the deflection were derived from the same source so that the system was synchronised and the sample was taken at the same point in each cycle. The deflecting coils were mounted on a bobbin around the neck of the tube, and rotation of this bobbin varied the instant at which the waveform was observed. An electrometer was used to measure the sample, but the system was not particularly satisfactory due to the finite de-ionisation time of the gas and various other sources of inaccuracy. In a second model, the beam of cathode rays passing through the slit was allowed to pass into a second tube, the deflection in the second tube being determined by the voltage at the instant of measurement. This was observed on a fluorescent screen of the usual sort.

The stroboscopic sampling principle came into its own again after World War II as a method for extending the frequency range over which the oscilloscope could operate [118,119].

References

[1]   Töpler A 1867 *Phil. Mag.* **33** 16–27
[2]   Hospitalier E 1901 *Bull. Soc. Int. des Électriciens* (Series 2) **1** 339–59
[3]   Report 1901 *Electrical World and Engineer* **38** 691
[4]   Beckit Burnie W 1897 *L'Éclairage Électrique* **13** 365–9
[5]   Beckit Burnie W 1897 *Electrician* **39** 849–55
[6]   Mikola S 1906 *Ann. Phys.* **20** 619–26
[7]   Joubert J 1880 *J. Physique* **9** 297–303
[8]   Golding E W 1935 *Electrical Measurements and Measuring Instruments* (London: Pitman) pp523 ff
[9]   Lenz E 1849 *Poggendorff's Ann.* **76** 494–523
[10] Lenz E 1854 *Poggendorff's Ann.* **92** 128–52
[11] Bowers B 1975 *Sir Charles Wheatstone* (London: HMSO) pp164ff
[12] Discussion 1891 *Trans. Am. IEE* **8** (May 21st) p393
[13] Thomas B F 1892 *Trans. Am. IEE* **9** 263–270
[14] Armagnat H 1897 *L'Éclairage Électrique* **12** 346–53
[15] Blondel A 1891 *La Lumière Électrique* **41** 401–8, 507–16
[16] Barbillon L 1904 *Manipulations et Études Électrotechniques* (Paris: Dunod) pp197ff
[17] Lamb C G 1906 *Alternating Currents* (London: Arnold) pp123ff
[18] Anon 1900 *Nature* **63** 142–5
[19] Ryan H J 1889–90 *Trans. Am. IEE* **7** 1–19
[20] Ryan H J and Merritt 1890 *La Lumière Électrique* **35** 233–240
[21] Feldmann C P 1894 *Wirkungsweise, Prüfung und Berechnung der Wechselstrom-Transformatoren* (Leipzig: Oscar Leiner) p391
[22] Torey and Walbridge 1890 *La Lumière Électrique* **38** 582–90
[23] Duncan 1891/2 *Trans. Am. IEE* **9** 179–91
[24] Duncan L 1892 *The Electrical Engineer* **10** 213–17
[25] Brackett 1906 *Electrical World and Engineer* **48** 1252
[26] Smith C F 1897 *Electrician* **39** 855–6
[27] Simmons H H 1912 *Electrical Engineering* (London: Cassell) p338
[28] Niethammer F 1900 *Elektrotech. Z.* **21** 309
[29] Searing L and Hoffmann S V 1889 *J. Franklin Inst.* **128** 93–101
[30] Bedell F *et al* 1893 *Trans. Am. IEE* **10** 497–518
[31] Anon 1900 *Electrical World and Engineer* **35** 161
[32] Fessenden R A 1896 *Electrical World (NY)* **28** 688–90
[33] Goldschmidt R 1902 *Elektrotech. Z.* **23** 496–7

[34] Goldschmidt R 1902 *The Electrical Engineer* **30** 117–118
[35] Krause R 1907 *Messungen an Elektrischen Maschinen* (Berlin: Springer) p124
[36] Ryan H J 1899 *Trans. Am. IEE* **16** 345–52
[37] Lamb C G *see* [17] p126
[38] Thomas B F *see* [13] p267
[39] *see* discussion following [36] pp353–60
[40] Mershon R D 1891 *Electrician* **27** 561
[41] Thomas B F *see* [13]
[42] Feldmann C P *see* [21] p390
[43] Hicks W M 1895 *Electrician* **34** 698–700
[44] Carpentier J 1903 *J. Physique* (Series 4) **2** 689–92
[45] Laws F A 1901 *Proc. Am. Acad. Arts and Sciences* **36** 321–4
[46] Fleming J A 1895 *Electrician* **34** 460–2
[47] Fleming J A 1901 *Handbook for the Electrical Laboratory and Testing Room* (Electrician) p398
[48] Fleming J A 1896 *The Electrical Engineer* **18** 73–6, 101–3, 129–32 etc.
[49] Robinson L T 1904 *US Patent Specification* 768,953
[50] Robinson L T 1905 *Trans. Am. IEE* **24** 185–214
[51] Report 1901 *Electrical World and Engineer* **37** 688–9
[52] Lyle T R 1903 *Phil. Mag.* (Series 6) **6** 549–59
[53] Laws F A *see* [45]
[54] Barr J M *et al* 1895 *Electrician* **35** 719–21
[55] Barr J M *et al* 1895 *L'Éclairage Électrique* **5** 171–5
[56] Report 1895 *Engineering* **60** 563
[57] Blondel A 1891 *La Lumière Électrique* **41** 507–16
[58] Report 1891 *Electrician* **27** 603–4
[59] Blondel A 1893 *La Lumière Électrique* **49** 501–8
[60] Duncan L 1892 *Trans. Am. IEE* **9** 179–91
[61] Duncan L 1891 *Electrician* **28** 61
[62] Duncan L 1892 *The Electrical Engineer* **10** 213–17
[63] Owens R B 1902 *Trans. Am. IEE* **19** 1123–9
[64] Laws F A *see* [45]
[65] Barr J M *et al see* [54] and [55]
[66] Carpentier J *see* [44]
[67] Rosa E B 1898 *GB Patent Specification* No 1872
[68] Rosa E B 1898 *Phys. Rev.* (Series 1) **6** 17–42
[69] Rosa E B 1897/8 *Electrician* **40** 126–8, 221–3, 318–21
[70] Rosa E B 1897 *Brit. Assoc. Report* (*Toronto*) *1897* pp571–4
[71] *Catalogues of James G Biddle Co.* (*Philadelphia*) 1899–1903 (in possession of Leeds and Northrup Ltd, Philadelphia)
[72] Laws F A 1938 *Electrical Measurements* 2nd edn (New York: McGraw-Hill) p644
[73] Robinson L T *see* [50] p192
[74] Catalogue 1905 *Leeds and Northrup Ltd*, (*Philadelphia*) p78
[75] Franke R 1899 *Elektrotech. Z.* **20** 802–7
[76] James R W 1901 *L'Éclairage Électrique* **29** 238–42
[77] Krause R *see* [35] p125

[78] Cambridge Scientific Instrument Co. 1912 The Brearly Curve Tracer (Leaflet No 196)

[79] Report of Phys. Soc. Exhibition 1910 *Engineer* **110** 666–8

[80] Report Brit. Assoc. Meeting 1895 *Engineering* **60** 563

[81] Lutoslawski M 1896 *Electrotech. Z.* **17** 211–13

[82] Callendar H L 1898 *Electrician* **41** 582–6

[83] Maddock A J 1956 *Proc. IEE* **103B** 617–32

[84] Drexler F 1896 *Z. Elektrotechnik* **14** 237

[85] Report 1896 *The Electrical Engineer* **18** 37

[86] Discussion 1903 *J. IEE* **33** 95

[87] Peukert W 1899 *Elektrotech. Z.* **20** 622–3

[88] Peukert W 1899 *L'Éclairage Électrique* **2** 108

[89] Hospitalier E 1903 *J. IEE* **33** 75–94

[90] Hospitalier E 1902 *Electrical Review* **50** 969–71, 1040–1 and 1903 **53** 1006–7

[91] Report 1901 *The Electrical Engineer* **28** 363

[92] Hospitalier E 1901 *Soc. Int. Électriciens Bull.* **1** 339–58

[93] Hospitalier E 1903 *Soc. Int. Électriciens Bull.* **3** 283–5

[94] Reeves R J D 1959 *Electronic Engineering* **31** 130–7, 204–12

[95] Simmons H H *see* [27] p339

[96] Rupp H 1900 *Elektrotech. Z.* **21** 820–2

[97] Wilke A 1895 *L'Éclairage Électrique* **4** 367–8

[98] Cardew P 1894 *Proc. R. Soc.* **56** 250–2

[99] Report 1895 *L'Éclairage Électrique* **2** 425–6

[100] Townsend F 1900 *Trans. Am. IEE* **17** 6–13

[101] Report 1900 *Electrical World and Engineer* (*NY*) **35** 161

[102] Gall D C 1932 *J. Sci. Instrum.* **9** 262–4

[103] Dannatt C *Metropolitan-Vickers Gazette* (no volume numbers used), August 1933 pp184–6

[104] Lyle T R 1903 *Phil. Mag.* (Series 6) **6** 549–59

[105] Stubbings G W 1929/30 *Commercial AC Measurements* (London: Chapman and Hall) 305–6

[106] Hubbard J C 1913 *Phys. Rev.* (Series 2) **1** 247–9

[107] Hay A 1894 *Electrician* **33** 425–6

[108] Grawinkel C and Strecker K 1891 *Electrician* **26** 459–61

[109] Grawinkel C and Strecker K 1891 *Elektrotech. Z.* **12** 6–7

[110] Wheatstone C 1843 *Phil. Trans. R. Soc.* Part II, pp303–327

[111] DuBois-Reymond E 1866 *R. Inst. Proc.* **4** 575–93

[112] Henry C 1894 *La Lumière Électrique* **51** 101–12

[113] Burdon-Sanderson J 1878 *J. Physiol.* **1** 196–212

[114] Bernstein J 1871 *Untersuchungen uber den Erregungsvorgang im Nerven- und Muskelsysteme* (Heidelberg)

[115] Burdon-Sanderson J 1879/80 *Proc. R. Soc.* **30** 383–7

[116] Report 1880 *Electrician* **5** 16–18

[117] Lübke E 1919 *Electrician* **83** 270–2

[118] Czech J 1965 *Oscilloscope Measuring Technique* (Eindhoven: Philips Technical Library) pp154ff

[119] Reeves R J D 1959 *Electronic Engineering* **31** (March) 130–7, (April) 204–12

4

Moving-Coil Oscillographs

Summary

Some early attempts to use the moving-coil galvanometer for the recording of waveforms will first be described. These were of very limited application; because of the inertia of the moving element, they were only able to cope with waveforms of very low frequency. The commercial frequencies of 50 or 60 Hz were beyond their capability to all intents and purposes.

Theoretical consideration will then be given to the fundamental problems associated with instruments of this sort, leading to the idea of the bifilar instrument conceived in France by André Blondel. His development of the principle, together with the work of William Duddell in Great Britain, will then be described.

Although the two people just mentioned dominated the field of bifilar oscillography, others found it convenient to develop their own instruments, either for their own immediate purposes or for commercial instruments, and these will be considered. Finally, the single-string oscillographs, which can be considered the ultimate simplification of the type, will be described.

4.1 Use of the moving-coil galvanometer

By the end of the nineteenth century, the moving-coil galvanometer was a well established method of measuring DC currents and voltages, having first been used by Sturgeon in 1836 and having been refined and developed by D'Arsonval and Deprez in 1882. It was comparatively easy to turn this into a recording meter which could make a permanent record of slowly varying currents. The pointer was simply made to bear lightly on the surface of a smoked rotating drum, or else it was equipped with a lightweight pen which drew a record on a moving strip of paper. Figure 4.1 is an example of this sort of thing, this particular instrument being due to M M Richard [1]. The moving-coil meter itself is situated in the horizontal box. It is equipped with a normal pointer and scale at the front but has a further pointer carrying a small inked pen which makes a trace on a band of paper wrapped around the drum. The drum is caused to rotate by means of an internal clockwork movement.

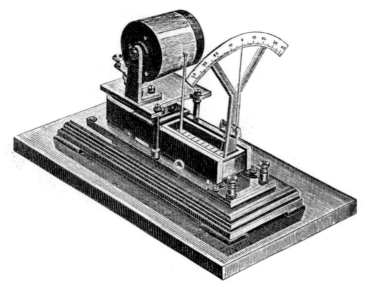

Figure 4.1 A moving-coil galvanometer adapted for the recording of slowly varying currents [1]. Courtesy: Gauthier-Villars, Paris.

Instruments of considerable sophistication were later developed for recording such slow variations in current; Callendar's recorder has already been mentioned in a previous chapter. Since the present work is more concerned with the registration of waveforms of mains frequency and above, we shall exclude detailed discussion of such recorders. However, readers who would like to see an example of the high degree of complexity and development achieved by instruments of this type may care to consult a paper written by R Shida in 1886 which described an instrument for studying earth currents in telegraph systems [2]. A series of thirty or forty closely spaced pins mounted on the moving-coil system were arranged to touch, one at a time, a fixed contact, these pins in turn each controlling an individual siphon pen recorder making a trace on a moving paper band. The current curve was delineated by the combined profile of all the traces.

It was quite a different matter when it came to using the moving-coil mechanism for determination of the waveforms of alternating currents and voltages at what were often referred to as 'commercial frequencies'—i.e. 50 or 60 Hz. The present chapter will be concerned with modifications to the design of moving-coil meters so that they would be able to cope with this relatively high frequency of alternation. We have already mentioned Elihu Thomson's attempt to use a moving-coil system (figure 2.45), and to some extent the material covered in that section overlaps the methods to be studied here since both are attempts to make mechanical systems vibrate at high frequency. Some of the theory which will be outlined is also applicable, with modifications, to diaphragm systems. One can appreciate instinctively that one of the limiting factors is the inertia of the moving parts. The pointer is particularly important here since its mass is situated some distance out from the axis of rotation. It was clearly a good idea to make this as light as possible and it became almost universal practice to employ a mirror and light beam as a 'weightless pointer'.

In 1892, G S Moler [3] attempted to modify an ordinary moving-coil meter in order to observe the waveform of the current produced by an alternator. One conductor in the rotor of the machine was connected via slip rings to the meter. The pointer was provided with a very light aluminium stylus which made a mark on a rotating smoked drum. The curves were then transferred from the drum to paper by damping the paper and rolling the drum over it. The frequency of the current was said to be '103 oscillations per second'. In the older literature, frequency was often specified as being so many 'vibrations per second', two vibrations constituting one cycle of modern terminology. It seems likely therefore that the frequency in this case was 51.5 Hz. The traces obtained were of very small amplitude indeed, and in the discussion which followed the presentation of Moler's paper to the American Institute of Electrical Engineers a Professor Crocker expressed the opinion that any differences which might exist between the various curves shown were so small as to be quite undetectable. However, another contributor to the discussion thought that the curves 'would bear magnification' and a further comment was made to the effect that a cylindrical lens would be useful for examination of the traces 'magnifying the ordinates of the curves more than the abscissae'. The rather limited success of the method can be inferred from these various comments.

C J Spencer [4] used a rather similar method in 1904 for observing the EMF in the rotor of an induction motor. A reduction gear was used to produce a well defined percentage slip in the machine and a simple moving-coil galvanometer with mirror cast a spot of light upon a moving film to trace out the waveform. H Becker, G Larmayer and G D Picard in 1889 [5] used the apparatus shown in figure 4.2 for recording the output voltage from a generator, and also for studying the primary and secondary currents of transformers. It consisted once again of an ordinary moving-coil galvanometer G, the field of which was provided by battery-energised coils. Voltage from the rotating machine H was connected to the moving coils through the terminals gg, and a sliding shunt S was provided in order to vary the sensitivity. An arc lamp D threw a beam of light on the mirror and thence onto a moving photographic strip M. Another beam of light was reflected from another mirror which was vibrated by an electrically maintained tuning fork F and thus a simultaneous calibration trace was also recorded. In a variant of the arrangement the arc lamp was energised by an induction coil, the primary current of which was interrupted at known frequency by a tuning fork. The trace was thus recorded as a series of spots of light, the time intervals between the spots being known.

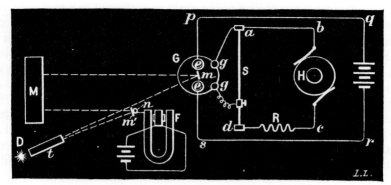

Figure 4.2   Apparatus used by Becker, Larmayer and Picard for determination of the waveform produced by an alternator [5].

This was again a straightforward attempt to use a moving-coil meter to trace an alternating waveform, but in order that it could operate successfully the speed of the generator had to be reduced to various values between 20 rpm and 300 rpm. All the curves obtained were sinusoidal in shape, and the authors could see no reason, therefore, why the waveform should not continue to be sinusoidal when the speed was raised to that normally employed. They attempted to carry out a mathematical study of the movement of the coil but their analysis was very limited in scope. However, even this was sufficient to show that when the applied frequency was high 'one would have only imperceptible vibration of the coil which would not allow interpretation of the results'.

Eric Gerard, the Belgian engineer from Liege, [6,7] was another experimenter who attempted to use a galvanometer for tracing the waveforms of alternating currents. A paper he presented in 1890 is largely concerned with the production of a dotted curve by means of an intermittent arc light as described above. The frequencies of the alternation seem to have been very low; the curve reproduced in the paper is of frequency about 2 Hz. His remarks indicate that he not only realised the importance of reducing the inertia of the moving system, but he may also have realised the advantages of making the system 'dead beat'—i.e. making it critically damped to avoid excessive oscillation. The significance of critical damping was already understood as regards DC galvanometers used for bridge measurements, but as we shall see later, this factor was also of importance when it came to using moving-coil instruments for waveform tracing.

4.2 Theoretical considerations

In order to appreciate fully the material of this chapter it is necessary to understand something of the mathematics describing a moving system of this sort. Consider the case of a moving coil, suspended on a torsion thread (or pivoted and restrained by a hair spring) and situated in a magnetic field which remains constant in strength as the coil rotates. Assume that the coil is carrying a current $i(t)$ which varies with time and which causes it to experience a deflecting torque $Gi(t)$. G is the so-called galvanometer constant which depends on the physical parameters of the system; the t in the bracket emphasises that the current i is a time-varying quantity. Let the deflection from the starting position at any time be designated $\theta(t)$. The torsion fibre or hair spring will exert a restoring torque upon the coil which will be proportional to the angle turned through; this torque will be designated $T\theta(t)$. The rotation of the coil will also be opposed by a viscous torque created by motion through the air or through any other fluid in which the coil may be immersed. In general, the faster the coil moves, the greater will be this frictional torque or 'damping' and it may be represented as $K\,\mathrm{d}\theta(t)/\mathrm{d}t$. Note that this assumption of proportionality may not be *precisely* correct, but it is generally considered to represent the situation to at least a reasonable approximation.

Knowing the forces at work upon the moving coil, we may now write the well known relationship

$$\text{Force} = \text{Mass} \times \text{Acceleration}$$

in its torsional form

$$\text{Torque} = \text{Moment of Inertia} \times \text{Angular Acceleration}$$

$$Gi(t) - T\theta(t) - K\frac{\mathrm{d}\theta(t)}{\mathrm{d}t} = P\frac{\mathrm{d}^2\theta(t)}{\mathrm{d}t^2} \tag{4.1}$$

where P is the moment of inertia of the moving coil. The negative signs precede those terms which represent forces opposing the motion of the coil. Rearranging the terms in this equation gives

$$P\frac{\mathrm{d}^2\theta(t)}{\mathrm{d}t^2} + K\frac{\mathrm{d}\theta(t)}{\mathrm{d}t} + T\theta(t) = Gi(t). \tag{4.2}$$

This is a second-order differential equation of the same type as that mentioned in Chapter 2 in connection with Kelvin's analysis of the Leyden jar discharge. If the current through the coil jumps suddenly from zero to a value I_f, then the solution of the equation can be shown to be of the general form [8]

$$\theta(t) = \theta_f\left(-\frac{m_2}{m_2 - m_1}\exp(m_1 t) + \frac{m_1}{m_2 - m_1}\exp(m_2 t)\right) + \theta_f \tag{4.3}$$

where θ_f is the final value of the deflection at which the coil settles, given by the equilibrium equation

$$GI_f = T\theta_f. \tag{4.4}$$

The values of m_1 and m_2 in the equation are

$$m_{1,2} = -\frac{K}{2P} \pm \left(\frac{K^2}{4P^2} - \frac{T}{P}\right)^{1/2}. \tag{4.5}$$

If $K^2/4P^2 > T/P$, m_1 and m_2 will be real and the deflection will rise smoothly to the final value θ_f in an exponential manner. On the other hand, if $K^2/4P^2 < T/P$ then m_1 and m_2 will be complex numbers of the form $(-a \pm jb)$ and the solution for $\theta(t)$ will be

$$\theta(t) = \theta_f - \theta_f \exp(-at)\left[\left(\frac{a^2+b^2}{b^2}\right)^{1/2} \sin\left(bt + \tan^{-1}\frac{b}{a}\right)\right] \tag{4.6}$$

where

$$a = \frac{K}{2P} \qquad b = \left(\frac{T}{P} - \frac{K^2}{4P^2}\right)^{1/2}.$$

This represents an oscillatory motion where $\theta(t)$ swings to and fro about the final value θ_f. The frequency of the oscillation will be given by

$$f_0 = \frac{1}{2\pi}\left(\frac{T}{P} - \frac{K^2}{4P^2}\right)^{1/2}. \tag{4.7}$$

This analysis has been well known for many years as we saw in an earlier chapter and has been used, for example, to optimise the conditions under which measurements are made with bridge circuits. An experimental circuit where the galvanometer has the second type of motion (equation (4.6)) is said to be 'underdamped' and is a great nuisance when making measurements because one has to wait a considerable time for the oscillatory movements to die away before the final reading θ_f can be taken. The other type of response is said to be 'overdamped' and this is almost as much of a nuisance, since the attainment of the deflection is long delayed and a wait is again necessary before the final deflection can be read. In practice, the optimum situation is achieved by adjusting the parameters of the system in such a way that the movement of the coil lies on the boundary between the overdamped and underdamped conditions. To achieve this $K^2/4P^2$ must equal T/P and under these conditions the movement is said to be 'critically damped' or 'dead beat'. Actually, if one carries out a very detailed analysis of the system, the best situation for speediest settling at the final deflection is to make the system just slightly underdamped. However, this is a counsel of perfection and in practice it suffices to provide approximately critical damping.

It should be noted that there is actually one term missing from the equation first derived (equation (4.1)). When a coil moves in a magnetic field it cuts the flux and an EMF is induced in it. Since this EMF tends to oppose the initial current which gave rise to it, it is often known as the 'back EMF'. This is proportional to the rate of cutting flux—i.e. to $d\theta(t)/dt$. If the inductance of the coil is small this can be accounted for by replacing the term $K d\theta(t)/dt$ by $(G^2/R + K)d\theta(t)/dt$ where R is the total resistance of the circuit in which the galvanometer is connected (including the resistance of the galvanometer coil itself). If the circuit resistance is large, little error will be incurred by neglecting this 'magnetic damping' and reverting to the form of equation (4.1).

Several attempts were made to extend this analysis to the case of a galvanometer which was carrying a current which was varying in magnitude and, in particular, to the problem of adjusting the various parameters to ensure that the deflection $\theta(t)$ was always directly proportional to the current $i(t)$ at every instant of time so that it could be employed to record faithfully the waveform of the current. In 1887 A Cornu [9] published an article giving an analysis of the problem and this was later taken up by his pupil A Blondel [10,11] who subsequently wrote a series of papers providing a highly detailed study of the behaviour of the galvanometer under various conditions. We cannot possibly reproduce all his conclusions here; let us instead observe a few simple facts. If, in the differential equation (4.2), we let P and K be so small as to be negligible compared with T and G, then we are left with the equation

$$T\theta(t) = Gi(t). \tag{4.8}$$

This simply means that $\theta(t)$, the deflection at any time, will be directly proportional to the current at that time, which is precisely what is required in a waveform-recording instrument. The implication of this is that the coil should be as small as possible so that it has a small moment of inertia P, and the damping (represented by K) should also be small. The suspension wire should be as taut as possible (T large : a small deflection produces a large restoring force) and the instrument should have a high sensitivity (G large: a small current produces a large deflecting force). To summarise, it needs a small coil on a taut suspension. This will have a very high natural frequency of vibration.

This was the approach which dominated thinking on the subject although, as we shall see later, it was not the only line of reasoning. Blondel studied in great detail the response of the moving system, a great part of his work being aimed at finding the amplitude of the galvanometer's vibration for an applied current of any frequency and also the phase difference between the current and the movement of the galvanometer. He showed that the galvanometer's performance was given by the formula

$$\frac{\text{Indicated amplitude}}{\text{True amplitude}} = \frac{1}{1 + (f/f_0)^2} \tag{4.9}$$

where f_0 is the natural resonant frequency of the galvanometer coil and f is the frequency of the applied voltage [12]. As a result of his studies he laid down the following criteria for a galvanometer to follow faithfully the variations of the applied current [13–15].

(i) The vibrating element (coil and mirror) should have a high natural frequency of oscillation compared with the frequency of the waveform studied. This follows from equation (4.9). A ratio of f_0/f of fifty times was advised; thirty times was considered acceptable.

(ii) The damping should be at the critical value. (It will be recalled that for this condition $K^2 = 4TP$, so this requirement can be reconciled to some extent with the need for large T and small P since they appear as a product in this expression.)

(iii) The inductance of the galvanometer coil should be small so as not to introduce a time lag into the measurements.

(iv) Induced 'eddy' currents in the moving system should be avoided.

(v) The sensitivity of the instrument should be as large as possible. In order to achieve this the number of turns on the coil should be as large as possible but this means an increase in the mass (and hence the moment of inertia) and so this requirement is in conflict with (i) and (iii) above to a considerable extent. The moment of inertia can be reduced by making the active sides of the coil as long as possible—i.e. by making the coil tall and narrow. The sensitivity can also be increased without resorting to an increased number of turns by making the magnetic field in which the coil moves as strong as possible

In a very concise summing up of all these results published in 1912 [16], J K A W Salomonson, Professor of Radiology at Amsterdam, considered the case of a system having a natural frequency of oscillation of 10000 Hz and showed that for an applied current of frequency 1000 Hz the amplitude of the instrument's response was reduced by only 1 % from the response to a current of much lower frequency. Even at an applied frequency of 3000 Hz the response was only 9 % down and the phase lags in both cases were negligible. He went on to show that with minor adjustments to the damping even better figures than this could be obtained. Thus it would appear that the recommended factors of fifty or thirty times mentioned above were unnecessarily large for all but the most accurate work.

In a paper written in 1894, Blondel [17] presented a method of correcting the curve drawn by a galvanometer in order to deduce the true current waveform. In other words, he corrected for the inaccuracies introduced by the non-ideal parameters of the vibrating system; as he phrased it 'allowing one to eliminate definitively every residual cause of error resulting from the inertia of the instrument and the self-inductance of the coils'. His method consisted of sampling the displayed curve at equal intervals of time and then using these sampled values to make estimates of the first and second derivatives of the curve. Addition of these in the correct proportions to the sample values themselves produced corrected sample values which represented the true curve of the applied current.

André Eugène Blondel figures prominently in the story of the development of waveform-recording instruments, and so it will be appropriate to say something of the history of this talented and prolific French physicist. He was born in 1863 at Chaumont in the Department of Haut-Marne. He studied at both the École Polytechnique and the École des Ponts et Chaussées in Paris. After qualifying, he was attached to the government department which had responsibility for lighthouses and navigation lights, and he remained closely connected with that department throughout his career. Ill health confined him to his room for twenty-seven years, although this in no way impaired his scientific work and the carrying out of his duties. As well as his work on oscillographs, which will be described shortly, he made important contributions to the science of photometry, to the study of synchronous motors, to radiowave propagation and the filings coherer, and also to radio direction finding. He was a recipient of both the Faraday Medal of the Institution of Electrical Engineers in London, and the Kelvin Medal, a prize established jointly by nineteen Institutions in Britain and the USA.

4.3 Blondel, Duddell and the bifilar oscillograph

It was Blondel who first suggested, in 1893, achieving the required operating conditions by reducing the coil of the galvanometer to its elemental form of a single loop of wire as shown in figure 4.3. The loop of wire is held under tension between the two poles of a magnet and the current under observation is passed up one side and down the other, one side experiencing a deflection forwards, the other backwards. A very tiny sliver of mirror fixed across the wires at their centres allows the deflection to be observed by means of a light beam [18,19]. Considerable magnification is possible by using a light beam in this way. For example, in his book 'Oscillographs' published in 1925, J T Irwin quotes some figures concerning the relative movements of the wires themselves and the light spot cast upon a screen. 'With the wires spaced 0.3 mm apart and a scale situated at a distance of 50 cm the distance moved by the projected spot will be 6,600 times that of the wires themselves' [20]. Various optical arrangements were employed for these systems, a common one being that shown in figure 4.4, originally suggested by Professor C V Boys [21−25]. A bright light is used to illuminate a vertical slit and the image of the slit is formed upon the oscillograph mirror with the aid of a plano-convex lens situated just in front of the loop. The reflected light also passes through this lens but before it is allowed to fall upon the screen or moving photographic film it is passed through a cylindrical lens which is curved in the vertical plane so that the vertical image of the slit is compressed down into a single intense spot.

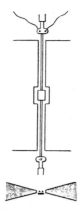

Figure 4.3 Blondel's original suggestion for a bifilar loop oscillograph (1893). The plan view (bottom) shows the magnet pole pieces which concentrate the field into the vicinity of the loop [18]. Courtesy: Gauthier-Villars, Paris.

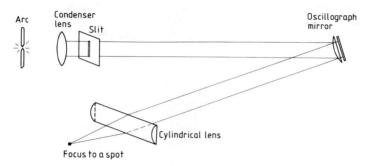

Figure 4.4 Cylindrical lens optical system for oscillographs; suggested by C V Boys.

In order to provide the necessary damping for dead beat operation the whole vibrator unit (i.e. loop and mirror) was immersed in oil. Paraffin, castor oil, glycerine and Canada balsam are frequently mentioned as suitable substances for this purpose [26,27] but many other oils, or mixtures of oils, were also used. The vibrating movement was situated in a bath of the oil, the bath being provided with an optical window to allow for observation of the mirror. Pole pieces which concentrated the magnetic field into the immediate vicinity of the wires were also set into the sides of the oil bath. It will be noticed from equation (4.7) that the expression for the natural resonant frequency of vibration f_0 includes the damping constant K and that any increase in this term results in a decrease in f_0. This immersion in an oil bath always reduced the frequency range over which the instrument could be operated.

Blondel, in France, developed this simple loop (or 'bifilar' oscilloscope as it was often called) into a useful laboratory instrument. He originally introduced it in the early 1890s but then concentrated his attention for a number of years on his moving-iron instrument (which will be described in the next chapter), returning to the bifilar vibrator around 1900. Some details of the 1900 instrument are shown in figure 4.5 [28]. This was a two-channel model having two vibrating elements mounted in a single housing. A third tiny fixed mirror m, mounted on the screw head between the other two, was provided to make a zero line on the screen or the photographic film. The wires were made of aluminium, silver, phosphor-bronze or silicon-bronze and they passed over the two ivory bridge pieces DD above the mirrors and another ivory piece E below. The upper-bridge pieces were mounted on a common pivot rod A and they could be tilted independently by means of screws situated behind the mounting so as to align the horizontal positions of the mirrors, thereby setting the zeros of the traces. The wires were fixed to terminals (a) at their lower ends whilst at the top they passed over the pulleys PP which were held in tension by springs inside the box B. The tension could be adjusted by means of the knurled knobs bb.

When in use the wires were contained in a box of damping oil with a window at the front, the whole assembly being inserted between the poles of a powerful electromagnet—figure 4.5(b). Wires from the terminals (a) dipped into cups of mercury so that electrical contact could be established with the wire loop. In some models, the pole pieces of the magnet itself, together with two bronze plates, front and back, formed the container for the oil. With aluminium wires of lengths 10 to 15 mm between supports it was said that a natural resonant frequency of between 10000 and 15000 vibrations per second (i.e. 5000 to 7500 Hz) could be obtained with a sensitivity such as to produce a light spot displacement of 600 to 800 mm per ampere on a screen at a distance of one metre. This frequency range was, of course, perfectly adequate for the study of mains voltages containing several harmonics. Greater sensitivity could be obtained if desired, but this was accompanied by a decrease in the resonant frequency. An increase in sensitivity of five times caused a decrease in resonant frequency to about 2000 Hz. Silver wires were used in these more sensitive instruments.

A slightly later model of the Blondel bifilar oscillograph, dating from about 1901, is shown in figure 4.6 [29]. The two bifilar elements are inserted into a box containing the oil and connection is established via the wires at the top. Figure 4.6(b) shows the box inserted between the poles of a magnet. Extension pole pieces are provided inside the box in order to concentrate the field as much as possible into the immediate vicinity of the moving elements. Figure 4.7 is a diagram of the complete apparatus used

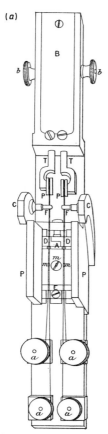

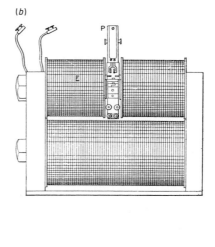

Fig. 4. — *Partie principale de l'oscillographe bifilaire de 1900.* — *a, a,* bornes-attaches des bandes; P, P, pièces polaires en fer, qui s'appliquent contre les pôles libres de l'électro-aimant; *c, c,* crochets-supports; F, F, bifilaires, en bandes d'aluminium ou de bronze; *m, m,* miroirs collés à cheval sur les bandes, et miroir de repère collé sur une tête de vis; E, appui inférieur des bandes en ivoire; D, D, appuis supérieurs en ivoire; A, pivot commun des appuis supérieurs (deux vis placées derrière l'appareil permettent d'obliquer l'un ou l'autre de ces appuis en les faisant tourner autour du pivot central A); *p, p,* poulies tendeuses; T, T, tiges tendues par des ressorts spiraux contenus dans la boîte B; *b, b,* attaches mobiles des extrémités supérieures des ressorts tendeurs.

Figure 4.5 Blondel's double bifilar oscillograph (circa 1900): (*a*), the two 'vibrator' elements; (*b*), overall view, showing elements in place between the poles of an electromagnet [13].

with these 'vibrators' (as they were sometimes called). The light source P_1 throws a beam of light onto the vibrating mirrors mounted in the electromagnet assembly on the right. The reflected light passes through the cylindrical lens C and thence from a rotating mirror M onto the ground-glass screen P where it can be traced by hand, or else it can be recorded on

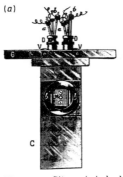

(a)

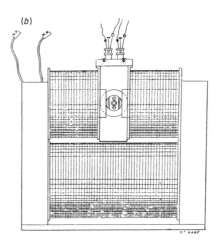

(b)

Fig. 11. — Pièce principale de
l'oscillographe bifilaire double
se plaçant entre les pôles
de l'aimant ou électro-aimant
(1901).

C, boîte en bronze avec pièces de
fer brasées à l'intérieur; G, plateau
supérieur; b, b, bifilaires; M, miroir de repère; V, vis tangentes pour
l'orientation; D, vis hélicoïdale pour
le réglage en hauteur; Q, tige à bouton moleté pour le réglage du plon-
gement des rayons; a et b, bornes
d'entrée et de sortie des courants.

Figure 4.6 1901 model of Blondel's bifilar oscillograph: (*a*), the two
elements contained in their oil box; (*b*), overall view showing the
elements between the poles of an electromagnet [22].

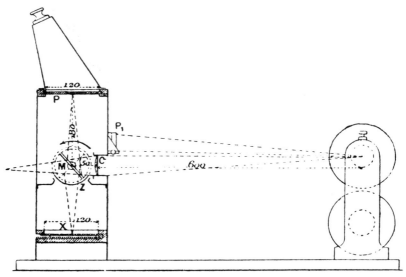

Figure 4.7 The optical arrangement of the Blondel oscillograph. The bifilar
loops and electromagnet are on the right; the rotating-mirror system for viewing or photographing the traces is on the left [18]. Courtesy: Gauthier-
Villars, Paris.

the photographic plate X situated in the bottom of the light-proof box. The rotating mirror is mounted on the shaft of a synchronous motor driven from the same supply as the waveform being observed, thereby producing stationary synchronised traces. Notice that since the observed current produces a horizontal movement of the beam and the time sweep is produced by the mirror rotating on a horizontal axis, the trace which appears on the screen is turned through 90° compared with that shown on a modern cathode ray oscillograph. A complete external view of the apparatus may be seen in figure 4.8. The box on the right contains the magnet and bifilar assemblies, and that on the left contains the rotating-mirror arrangement (often called a 'synchroscope'), the viewing screen, and a photographic slide arrangement. The two units are joined together by light-tight expanding bellows. The large box on the left, which has the general appearance of having been added on as an afterthought, contains an arc lamp.

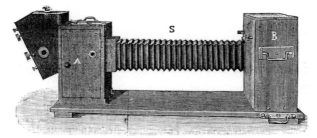

Fig. 17. — Vue d'ensemble de l'oscillographe bifilaire à électro-aimant pour laboratoire (1901).
B, caisse contenant l'électro-aimant; S, soufflet en toile noire; A, caisse formant chambre noire et contenant le synchronoscope et l'écran photographique ; L, projecteur.

Figure 4.8 Complete view of the Blondel laboratory oscillograph [22].

In a paper written in 1897, Blondel described the various uses to which his oscillograph could be put [30]. These included the simultaneous tracing of V and I curves using one bifilar vibrator as a voltmeter and the other as an ammeter, and also the plotting of V versus I curves (as opposed to plotting them as functions of time). This was achieved by the use of two vibrators situated at 90° to one another, the light beam being reflected from one mirror to the other and then onto the screen. Displays of this sort were sometimes known as 'cyclograms'. A method is also described for measuring rapidly-varying resistances such as, for example, the resistance of an incandescent lamp as it cools after the current is switched off.

The bifilar principle was taken up by W D Duddell in Great Britain [31–33], in the first instance, to assist his experiments on the electric arc. He made several improvements to the design of the loop, and by careful engineering and construction produced instruments of high quality and great usefulness which surprisingly enough, for reasons which will appear later, are still occasionally used today for special purposes. He and E W

Marchant first tried to carry out their arc experiments using a Joubert contact method but found it a very laborious process:

> After working with it for nearly three months we came to the conclusion that our investigations would never be completed unless some quicker and more accurate method was devised for obtaining the curves ... one of us therefore considered the subject of curve tracers and oscillographs, and more especially the methods which had been suggested by Monsieur Blondel and the methods on which experiments had been carried out by Professor Ayrton and Mr Mather.

The work of Ayrton and Mather was done in the years 1895–6 but, as far as the present author has been able to ascertain, they do not appear to have published the results of their experiments, and in a paper which they wrote to the *Philosophical Magazine* a few years later, they describe a Duddell instrument but modestly make no mention of their own efforts [34]. There is some evidence to show, however, that they demonstrated their apparatus at a British Association meeting [35]. It is perhaps significant that in a paper to the *Electrician* in 1897 [36] Duddell refers to their work as being 'not yet published'. In the concluding section of that paper he wrote '... the writer wishes to thank Professor Ayrton and Mr T Mather and the other teachers at the Central Technical College for much valuable advice and assistance and also for the use of the apparatus to carry out the experiments; he also wishes to acknowledge his indebtedness to Mr Marchant, his fellow student, for helping him with all the experiments'. Both Duddell and Marchant became, in their turn, Presidents of the Institution of Electrical Engineers. Marchant was, for a time, superintendent of Lord Blythswood's laboratory which was mentioned in an earlier chapter.

The Ayrton and Mather apparatus is now in the Science Museum in London (figure 4.9). It consists of a loop of wire held under tension as in the Blondel type, but the movement was observed by having a small bar connecting the two wires (replacing the mirror in the Blondel version) to which was attached a pig's bristle which made a trace upon a narrow smoked glass plate pulled by hand through a guide beneath it. As far as can be judged from the construction of the apparatus, the tension in the wire loop cannot have been very high and it seems unlikely that it could have been used with any great success to record mains frequency waveforms. However, it was clearly this crude device which started Duddell upon the construction and development of his own highly successful instrument.

Duddell's oscillograph is shown in diagrammatic form in figure 4.10; the actual instrument can be seen in figure 4.11. A great difficulty with Blondel's original bifilar vibrator had been in obtaining sufficient damping to produce the correct operating conditions. One of Duddell's major improvements was to use a flat phosphor-bronze strip in place of the simple wire loop. This presented a larger surface area to the oil and hence a greater

Figure 4.9 Ayrton and Mather's early bifilar oscillograph, now at the Science Museum, South Kensington. A long narrow strip of smoked glass rested on the roller guides and was moved by hand. N.B. The small bar bearing the bristle which marked the glass plate is now missing. Photograph: Science Museum (395/85).

resistance to movement and greater damping. In the very earliest instruments the strips were contained in a simple oil-chamber (shown dotted in figure 4.10), but in his later models the two halves of the strip were separated by a thin iron partition so that each part moved in its own very

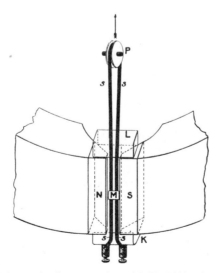

Figure 4.10 Schematic diagram of Duddell's bifilar loop oscillograph [31].

narrow chamber. The oscillograph shown in figure 4.11 was provided with two loops for two-channel use. The two loops passed over their individual pulleys. These pulleys were connected by a wire which passed over a third pulley and the tension was adjusted by the screwed rod at the top of the instrument, its value being measured by means of a spring balance. To the right of the central window which allowed observation of the mirrors can be seen a tube which was used to introduce the damping fluid into the chamber formed by the pole pieces of the magnet, a backing plate and the lens plate at the front. This tube could also be used as a sighting glass to determine the level of the oil; a tap situated at the bottom of the tube was used to empty the chamber prior to dismantling the vibrating elements should the need arise. The magnetic field was provided by the toroidal coils which are such a prominent and characteristic feature of the instrument.

Figure 4.11   Duddell's bifilar oscillograph (1899 model) [31].

Duddell had worked his way up from the ranks of the engineering profession. Before going to South Kensington to study under Ayrton and Mather, he had served his apprenticeship with a firm in Colchester. He was a highly skilled craftsman in his own right. In a foreword to a commemorative booklet published by the Cambridge Instrument Co. Ltd in 1950, [37], Professor Marchant recalled the meticulous attention to detail which was such a characteristic of Duddell's approach.

> The next problem was to find a mirror that could be attached to the strips, to indicate their deflection. The mirror had to be as light as possible and the glass surface absolutely flat and even, to give regular reflection. The thinnest glass we could think of was the cover glass for a microscope slide. The cover glasses were tested for flatness and equal thinness over their whole area by examining the interference fringes produced by a monochromatic sodium light, a method now universally used by the makers of flat mirrors and lenses; but, at that time, a novel means of testing. Having found a suitable glass, it was next silvered, and Duddell took endless trouble to find the best and most adherent silvering solution to employ; then the mirror had to be attached to the strips and this again was carefully thought out. If the mirror was stuck to the strips by resin, it would almost inevitably be twisted and distorted in the process, so four tiny dots of resin were first put on the strip, and the glass carefully positioned on them. In this way a perfectly flat mirror was obtained and a definite spot observed when a beam of light was reflected from the surface. ... Such details as the choice of damping oil which had nearly the same refractive index as glass were carefully considered, and use of a mixture of castor oil and clock oil provided the immediate solution.

It was important to ensure that the viscosity of the damping oil was such that the oscillations were critically damped. To this end, the oscillograph was first used to observe a waveform consisting of a periodically interrupted current. The DC current in the magnetising coil was then switched on and this gradually caused the temperature to rise, thereby decreasing the viscosity of the oil. The deflection was observed closely with the rotating-mirror arrangement, which will be described later, and the warming up process was allowed to continue until the observed waveform just began to overshoot due to underdamping. It was then allowed to cool slightly so that precisely dead beat operation was achieved. The temperature was then noted by means of a thermometer set into the oil chamber and thereafter the oil was always brought up to this temperature before use.

Figure 4.12 shows the optical arrangements used for a three-channel instrument, containing a single and a double instrument working together. Figure 4.13 shows the associated electrical circuitry. The double instrument O_2 has mirrors m_2 and m_3 on its loops and a further fixed mirror m_4 to produce a zero line on the trace. The single instrument O_1 has its beam superimposed on the others by means of the prism p. The lantern L illuminates a vertical slit and the Boys' cylindrical lens C is used to concentrate the spots as previously described. The trace could be recorded photographically by allowing a plate to fall through a tall vertical slide S.

It was estimated that by the time the plate reached the horizontal slot upon which the light spots fell it was travelling at a speed of $640\ cm\,s^{-1}$.

A special brake arrangement pressing on the back of the plate was used to decelerate it safely after exposure. Light-tight bags were used to load the plate at the top and unload it from the bottom after use.

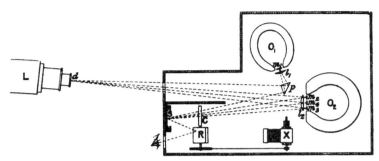

Figure 4.12 Optical   arrangement   for   Duddell's   three-channel oscillograph. O_1 has a single vibrator element; O_2 has two vibrators and a stationary mirror m_4 to provide an *x*-axis trace [31].

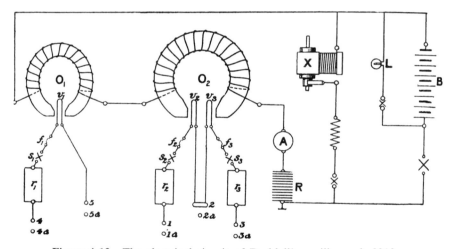

Figure 4.13 The electrical circuit of Duddell's oscillograph [31].

If required, the trace could also be viewed continuously through the rotating mirror R driven by the motor X. A white screen was provided at S and this was observed through the mirror. The waveforms could thus be monitored throughout an experiment except for the very short interval when the falling plate obscured the white screen. The speed of the DC motor rotating the mirror was adjusted by a series resistor. Since it was not synchronised to the observed waveform in any way it must have been rather difficult to hold a steady trace on the screen. The source of illumination was an arc lamp contained in the lantern L of figure 4.12. The lamp in figure 4.13, also rather confusingly marked L, was in fact a small bulb mounted inside the optical box which allowed one to read the thermometers mounted on the oscillographs. Each vibrating loop was connected in series with a fuse, a switch and a resistance box to vary the sensitivity, which was such as to produce a deflection of 1 mm on the screen for a current of 0.005 A. The undamped resonant frequency of the loops was 2700 Hz; this would have been lowered somewhat by the presence of the damping oil of course.

Duddell, with Marchant his collaborator, verified the accuracy of the oscillograph by photographing a 'peaky' waveform produced by an alternator (or dynamo as they called it) and comparing the trace with the results obtained by a Joubert contact measurement. As they commented,

> The practical identity of the points on the oscillographic curves with those obtained by the Joubert method is very strong, and where small differences can be detected we are unable to say which was the more accurate; for the speed of the dynamo was not, of course, absolutely constant; and it must be remembered that the oscillograph points have reference to time, while the Joubert points have reference to the angular position of the moving parts of the dynamo.

In other words, the setting of the moving contact in the Joubert method was subject to errors and thus one could not rely on its absolute accuracy.

Duddell went on to develop this instrument under the auspices of the Cambridge Scientific Inst. Co. Indeed, for the next thirty years or so they were produced in an almost bewildering variety of forms. The Science Museum possesses several versions of the early models. The earliest were clearly more-or-less experimental in nature, being mounted on fairly crude wooden bases. Duddell's very first instrument of all is shown in figure 4.14; he is said to have produced over a hundred experimental models altogether. As time went on these were replaced by a version having a cast metal base, this being indicative of the fact that they were beginning to be commercially produced. They were being improved all the time, and very soon the Cambridge Company were producing models having a resonant frequency of 10 000 Hz. It is said that for many years Duddell insisted upon inspecting every instrument personally before it was sold. A version dating from 1900 may be seen in figure 4.15 [38,39]. A lamp and condensing lens were placed outside the case to the right (not shown here). The light beam entered through a vertical slit cut in the side of the box and was reflected from the moving mirrors of the oscillograph, and thence through the cylindrical lens, images of the spots produced being cast upwards onto the curved ground-glass screen by an oscillating mirror.

The oscillating mirror system can be seen in greater detail in figure 4.16 [40]. It consisted of a synchronous motor (moving-iron attraction type) driven by a supply of the same frequency as that of the waveform

Figure 4.14 Duddell's first experimental instrument, now at the Science Museum, South Kensington. Photograph: Science Museum (119/50).

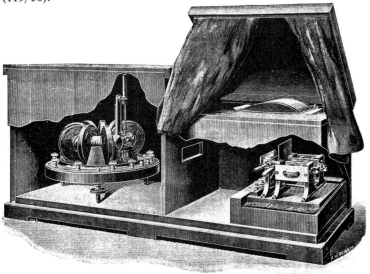

Figure 4.15 A complete Duddell oscillograph outfit (circa 1900) [38].

Figure 4.16   The oscillating mirror system (also seen at the right-hand side of figure 4.15) [40]. Courtesy: Cambridge Instruments Ltd.

being observed. A cam was mounted on the motor shaft and this was in contact with a follower attached to a frame carrying a thin strip of mirror. This can be seen in figure 4.16 immediately behind the cylindrical lens at the front of the apparatus. The profile of the cam was such that it produced a linear displacement with time of the spot of light across the curved screen. At the end of the sweep, the cam caused the mirror to return rapidly to its starting position, and during this 'flyback' period an occluding shutter, also fixed on the motor shaft, interrupted the light beam. The circular leaf of this shutter can be seen protruding above the top of the mirror. The sweep was arranged to occupy about 1.5 periods of the waveform under observation, the remaining half period being allowed for flyback. Various series/parallel arrangements of the motor coils were available to suit supplies of different voltages and frequencies. A brass disc was also provided to increase the inertia of the rotating armature when used with supplies of frequencies lower than 40 Hz. A lever at the front of the unit could be operated to disengage the rocking mirror while the motor was being brought up to synchronous speed. A permanent record of the trace could be made by laying a piece of tracing paper over the screen and tracing it by hand. The lid of the box was supplied with curtains to assist in viewing the trace which was typically about 6 cm peak to peak in amplitude by 8 cm length. A piece of photographic paper could also be laid across the screen to register a permanent image of the trace.

The loops of wire in the oscillograph itself were made of phosphor-bronze 7 mil wide by 0.3 mil thick. Each leg of a loop moved in a gap with 1.5 mil clearance on each side (1 mil = 10^{-3} in). A resonant frequency of 10 000 Hz was attainable. Fuses were provided to protect the delicate loops in the event of overload.

The Duddell oscillographs described thus far all incorporated powerful electromagnets to produce the strong magnetic field required. They were therefore rather bulky instruments. A more portable form using a horseshoe permanent magnet was devised in 1902/3 [41,42]. Single and double versions were available, the single model being shown in figure 4.17. These were of rather more modest specifications than the electromagnet versions and were intended as a cheap, easily transportable model for use in schools, factories and in the field. The resonant frequency of these movements was up to 5000 Hz. The complete case in which they were sold can be seen in figure 4.18. No separate arc lamp was required for illumination; the case contained an incandescent lamp which could be energised from the supply under observation or from any other convenient source. A rheostat was incorporated to control the brightness of the lamp and the case also contained a simple rotating mirror to enable the trace to be observed. This was turned by hand using the handle seen on the right-hand side of the

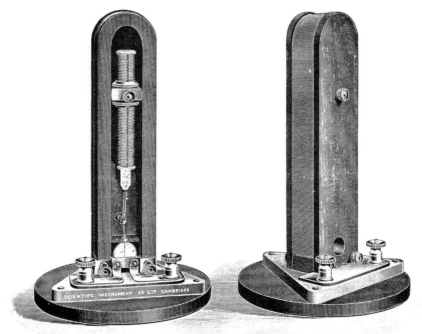

Figure 4.17 Duddell single-loop permanent magnet oscillograph [41]. Courtesy: Cambridge Instruments Ltd.

case. Rubber eye caps were provided to exclude extraneous light when observations were being made. Arrangements were available for the observation of voltage or current waveforms. It was advised that for voltages above a few hundred volts or so a transformer should be used to reduce it to a safe level.

Some more detailed figures specifying the performance of this model are:
resonant frequency 4000 Hz with tension of 1 oz;
sensitivity 200 mm per ampere at a distance of 25 cm;
resistance of the strips 4 ohms;
resistance of strips with fuses and connecting wires 14 ohms.

It was also arranged that the damping oil supplied, which was contained in the U-shaped cavity of the magnet, would have the correct viscosity at room temperature.

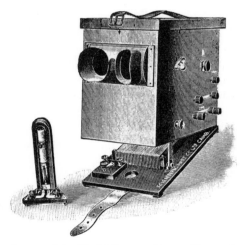

Figure 4.18 A complete Duddell portable oscillograph outfit [41]. Courtesy: Cambridge Instruments Ltd.

To give some idea of the general conditions of use and method of operation we may quote at length from a contemporary account [43]. The circuit diagram of the various connections is shown in figure 4.19; the large circles A to E represent the terminals on the instrument.

> The instrument can be used to show either P.D. or current waveforms of the circuit under examination. To investigate P.D. waveforms the terminals B and C are connected directly to the poles of, say, a 100 or 110 volt circuit. This can be done conveniently by putting the adapter supplied with the instrument into an ordinary lamp holder. The terminals D and E are connected to B and C respectively by means of the copper strips which are provided, so that the lamp is lighted from the same circuit. As will be seen from the diagram, there is a non-inductive resistance of 1,000 ohms permanently connected in

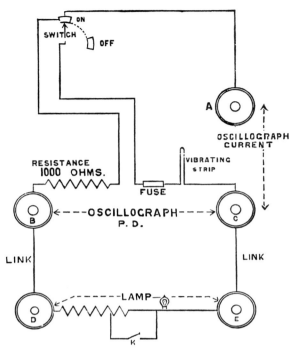

Figure 4.19 Circuit connections for portable oscillograph [41].
Courtesy: Cambridge Instruments Ltd.

series with the vibrating strip of the oscillograph, so that the current through
it is about 0.1 of an ampere, which gives an amplitude of about 25 mm on
each side of zero. There is also a resistance of about 90 ohms in series with
the Ediswan 'Miniature' 20-volt lamp. The key K is arranged to short circuit
a portion of this resistance thus increasing the brightness of the lamp when
the curves are actually under examination.

For voltages up to three or four hundred, resistances can be used in series
with the instrument. With higher voltages it is advisable that they should be
transformed down by means of a small transformer having a closed iron circuit and a small leakage. In this latter case, it is advisable to disconnect the
lamp terminals D, E from B, C and to connect them to an independent 100
volt circuit and to earth one of the terminals B, C as it is not safe to use the
instrument if it is more than a few hundred volts above earth. Another
method is to use a number of incandescent lamps in series as a potential
divider, the instrument shunting one of the lamps, which must be connected
to earth. It is not advisable to use this method for voltages above 2,000 or
3,000 volts.

To investigate current waveforms the terminals A and C are connected to
the potential terminals of a suitable low-resistance shunt in the main circuit.
This shunt should have a resistance so that at the maximum current there is

a P.D. across it of about 1.4 volts. In this case the lamp must be lit separately from a 100 volt circuit. For both these investigations, lamps having strong thick filaments should be used

The passage just quoted seems to imply that the instrument was not really suitable for use with voltages higher than 3 kV. However, the Cambridge Instruments catalogue for 1903 put quite a different slant upon the matter:

> In experiments on high voltage circuits, such as two- and three-phase transmission circuits, it is often impossible to arrange the connections so that the oscillograph is at earth potential. In these cases it is often necessary to experiment with 5,000, 10,000 or even higher voltages between the oscillograph and earth. For this purpose, the permanent magnet oscillograph is eminently suited as it can be easily insulated; and owing to its small size it has no appreciable electrostatic capacity, the possession of which might introduce errors under some conditions. A small ebonite table is made to carry the instrument for voltages up to 10,000 volts, and has three screwed ebonite legs which act as levelling screws.

The point was, of course, that with the *electro*magnet type, if the vibrating loop were at high potential relative to earth there would be considerable risk of the insulation breaking down—bearing in mind the close spacing of the loop in the air gap. The permanent magnet type, requiring no magnetising supply, could be allowed to 'float' at the high supply voltage. Special insulating bushes were necessary to make connection to the instrument inside the viewing box but provided that sensible precautions were taken, waveforms could thus be viewed with safety.

Duddell oscillographs were sold equipped with various means for recording the waveforms [44]. Figure 4.20 shows an outfit equipped with a falling-plate arrangement. The lid of the dark box is propped open against the photographic slide (on the left). The motor at the bottom of the box rotated a four-sided mirror drum by means of a belt and pulleys, and this cast the trace onto the viewing screen on the top of the box (not visible in this photograph due to the raised flap). When a suitable wave trace was seen, the photographic slide was released down the chute, a speed of $400 \, \text{cm s}^{-1}$ at exposure being attained. The outfit could be used with a continuous-motion cinefilm transport arrangement so that longer traces could be obtained. The film camera was inserted between the lantern and the main oscillograph box. A small external motor supplied the drive, and the motion of the film past the exposure slot was controlled by a gear-driven sprocket wheel. A brake and jockey-pulley system allowed the operator to let the main driving belt slip or to take up the drive as required; thus the film could be stopped and started easily with the minimum of wastage.

A recording drum system was also available as an optional extra. A strip of photographic film was wrapped, sensitive side out, around the drum and an exposing shutter could be set so as to expose any desired length of film from 10 cm to 40 cm. A range of film velocities up to 600 cm s^{-1} could be obtained and a contact device was also included to act as a trigger, initiating the phenomenon which it was desired to study. J T Morris [45], writing in the *Electrician* in 1907 considered this method 'clumsy and expensive' and chose instead to devise his own method in which a falling photographic plate cut through a fine thread at the appropriate point in its fall thereby releasing a triggered contact. He experienced considerable trouble with contact bounce but eventually found a type of knife blade and spring contact which was virtually bounce free.

Figure 4.20   Duddell oscillograph outfit (circa 1903) with falling-plate recording arrangement. The plate slide is on the left of the box, just behind the raised flap [41]. Courtesy: Cambridge Instruments Ltd.

The need was often felt for projecting the trace upon a large screen for the benefit of an audience. The arrangement of figure 4.21 was available for this purpose [46]. For normal laboratory use the rocking-mirror system, driven by its own motor, was placed inside the tracing box seen to the right, near the arc lamp. The trace was reflected upwards onto the curved screen in the usual way. For projection, the rocking-mirror unit was taken out of this box and placed instead in the box on the left. The trace was reflected upwards as before, but now the large inclined mirror projected it forwards onto the distant screen. A somewhat more robust version of the rocking-mirror unit was also manufactured. This was driven by a two-phase syn-chronous motor supplied from a phase-splitting unit. This would no doubt have provided a stronger, more positive drive than the simple reluctance motor previously described. It was also provided with a set of Joubert contacts so that point-to-point measurements could be performed if required.

The Double Projection Oscillograph.

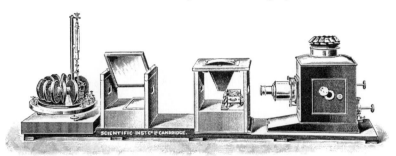

Figure 4.21   Duddell oscillograph with adapter for projection [41]. Courtesy: Cambridge Instruments Ltd.

There was one great difficulty with this projection type of apparatus. The mirrors of the normal types of oscillograph described so far were so small (about 3 mm × 1 mm on the original model, reduced to 1 mm × 0.3 mm on the high-frequency commercial models) that it was very difficult to throw sufficient light onto them to produce a reasonably bright projected trace. To improve this, a special projection model was available having a larger mirror [47]. This would project traces up to a distance of 6 ft, but owing to the increased inertia of the vibrating system its natural frequency was reduced to 2000 Hz and it was five times less sensitive than the normal 'high-frequency' model [48]. When used for projection, the vertical slit in front of the illuminating lamp was replaced by a circular aperture and the Boys cylindrical lens was removed [49].

A major disadvantage with the Duddell bifilar oscillograph was the great difficulty of repairing or replacing the moving element should it be accidentally burnt out or damaged in any way. By about 1906, a new simplified form of element had been introduced—figure 4.22. The main frame is made of brass, the wire loop carrying the mirror runs down from one of the brass clamping plates at the top, under the pulley fixed to a tensioning spring at the bottom and back to the second clamping plate at the top. Two pole pieces of soft iron are situated on either side of the loop, and the narrow channel in which it moves is further divided by a thin iron partition. To quote from one contemporary account [50] 'As each vibrating system is made as simple as possible no special skill is required to repair it, any

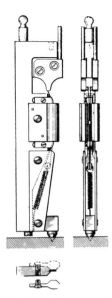

Figure 4.22 Simplified Duddell vibrator which eased the problem of replacing a broken loop [46].

reasonably skilled experimenter being able to re-suspend it completely in about a couple of hours. Any one who has endeavoured to re-suspend vibrators of the double, high-frequency type will realise what a great improvement Mr Duddell has introduced in this new simple vibrator'. Once again, this new vibrator was sold in two versions, one for high-frequency work and another for projection purposes. They were designed in such a way that they could be plugged in and out fairly easily and the redesigned electromagnet and housing to receive them can be seen in figure 4.23. They could also be used in a new version of the permanent magnet instrument shown in figure 4.24. Two vibrators could be housed side by side in the cylindrical oil bath (although only one is shown in position here), and the permanent magnet was fitted with extension pole pieces which projected into the oil bath in order to concentrate the flux near the elements. This version was particularly suitable for high-voltage work and a complete high-voltage oscillograph is shown in figure 4.25. The unit of figure 4.24 was housed within the porcelain dome seen in 'ghost' form at the right and voltages up to 10 kV could be studied using series resistors wound on porcelain tubes mounted on wooden frames and specially arranged so as to minimise inductive and capacitive effects. Even higher voltages could be handled using resistance wire woven into asbestos mats and immersed in tanks of oil. Note that the small box seen on the left of figure 4.25 is the cinefilm unit mentioned earlier.

Figure 4.23 Electromagnet and housing for the vibrator of figure 4.22 [40]. Courtesy: Cambridge Instruments Ltd.

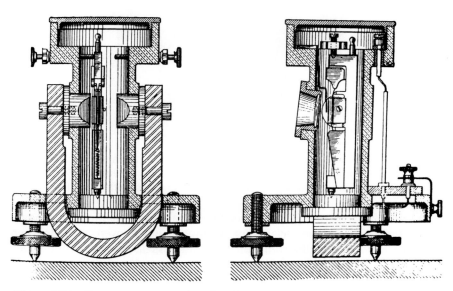

Figure 4.24 Permanent magnet housing for vibrators of the type shown in figure 4.22 [46].

The Duddell oscillograph was a highly successful instrument which was continually developed and improved over many years, and it is occasionally found in use even today. One might have imagined that it would have been entirely superseded by the cathode ray oscilloscope, but it does have two advantages which make it suitable for certain specific applications. It is relatively easy to make multichannel measurements with good interchannel isolation, and it is possible to use it for high-voltage applications, the system providing a high degree of insulation as previously described.

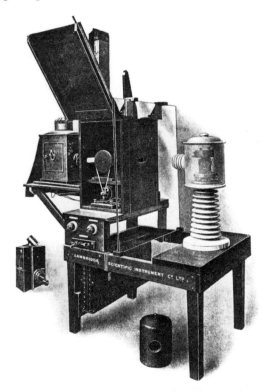

Figure 4.25 Duddell oscillograph used for high-tension work. The smaller item on the left is the camera for producing a continuous trace on 35 mm cine film [40]. Courtesy: Cambridge Instruments Ltd.

As time went by, the vibrators were reduced in size, making the production of multichannel models easier. Complete outfits were sold, tailored more or less individually to customers' requirements; a typical one which featured in the Cambridge Instruments Catalogue for 1932 is shown in figure 4.26. The vibrating elements used at this time had a resonant frequency in air of 12 500 Hz, and a sensitivity such that a 60 mA current gave a deflection of 20 mm on a screen at a distance of 60 cm. Many of these later outfits were equipped with quite sophisticated mechanical exposure and triggering systems. Perhaps we should also mention here that the bifilar loop principle was adapted in the late 1920s for the production of variable-width sound track on cinema films, RCA in the United States being one of the main companies involved in this development [51,52].

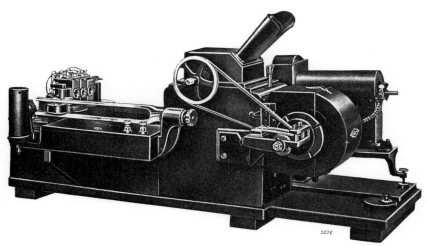

Figure 4.26 A typical complete Duddell oscillograph outfit which featured in the 1932 catalogue of the Cambridge Scientific Instrument Co. [94]. Courtesy: Cambridge Instruments Ltd.

It was generally accepted that the original idea for the bifilar form of the oscillograph had been due to Monsieur Blondel and that he had performed the detailed mathematical analysis which established the optimum operating conditions for such vibrators. However it was also widely felt that great credit was due to Duddell who had brought his considerable engineering skills to the development of the instrument. In particular, his introduction of the flat-strip form of loop in place of the simple wire was held to be a most important contribution although, oddly enough, in 1901 Blondel was also granted a patent [53] which included the use of two parallel bands, as opposed to one band in the form of a loop which had been specified in Duddell's patent. A comment made by David A Ramsay in 1906 is typical of many to be found in the literature of the time [54]:

> The original idea of the instrument is due to M. Blondel who pointed out in 1893 the principles on which such an instrument should be designed, but to Mr Duddell is due the credit of developing the instrument into its present form as a practical commercial instrument.

William Duddell who, by all accounts, was a most agreeable and courteous man, referred in his original IEE paper to 'the methods which had been suggested by M. Blondel' [55]. In a slightly earlier article in the 'Electrician' he stated quite clearly that '...the best results have been obtained by M. Blondels's oscillograph or some of its modifications' and went on to indicate that it was Blondel who first suggested the use of the bifilar loop and that his (Duddell's) own instruments were based on Blondel's idea [56].

Blondel, for his part, acknowledged the undoubted contribution made by Duddell [57]:

> The practical perfection of the bifilar oscillograph was actively undertaken in England in 1897 and 1898 by Mr Duddell who much improved the sensitivity and usefulness of the arrangement. Starting, as he has courteously recognised, from my description of 1893, that author has perfected the bifilar apparatus in several important details ... he was, above all, happily inspired in replacing my copper wire by phosphor-bronze bands (a configuration imitating the suspension of Ayrton and Mather galvanometers) by which he could reduce sufficiently the cross section to obtain good sensitivities with characterising frequencies of 3,000 then 10,000 cycles.

However, in reading the literature of the period, generally, one somehow gains the overall impression that Blondel was very touchy about not receiving due credit for the invention of the instrument. It was, after all, *his* idea which had been taken up by another worker and brought to a high degree of commercial development and he must have felt overshadowed by the growing success of Duddell and his collaborators, the Cambridge Scientific Instrument Company. His discontent boiled over in 1906 when David A Ramsay wrote the article in the *Electrician* which has been referred to earlier and from which a section was quoted above. In this article, Ramsay had referred in all innocence to the 'Duddell oscillographs'. Blondel immediately wrote a Note to the Editor, quoting the relevant references, pointing out that it was *his* oscillograph, not Duddell's and that in all fairness it ought to be called, if not the 'Blondel oscillograph' then at the very least the 'Blondel–Duddell oscillograph' [58]. He also claimed to have invented the word 'oscillograph'. The letter was very polite to Duddell himself; his complaint was that 'the makers of the Duddell oscillograph, probably for commercial reasons, have ignored my name in connection with the apparatus'. Duddell wrote a letter, slightly acid in tone, in the following issue of the *Electrician* defending his position [59]. For example, he referred to 'my 1897 paper in which I adopted the word "oscillograph" to do honour to M. Blondel's work, a word which I should not otherwise have inflicted upon the English language'. With reference to the numerous theoretical papers of Blondel he wrote 'I have not so far inflicted my mathematics on engineers'—which can perhaps be interpreted as a sly sideways swipe at the long turgid analyses which had been published by Blondel. He also suggested that since, even in France, manufacturers had adopted the flat-strip form of construction, Blondel should instruct *his* manufacturers to call their instruments 'Blondel–Duddell' oscillographs as well. He indicated, a little sarcastically, that the 'D'Arsonval' moving-coil galvanometer which Blondel had mentioned in his letter, should properly have been referred to as the 'Thomson–Maxwell–D'Arsonval' galvanometer to be consistent with the general argument. Duddell very sensibly concluded his letter by saying that 'when two people set out to attain the same end, in this case good records of waveforms, they are likely to work along parallel lines'.

That Blondel was very sensitive on the subject can also be illustrated by another little episode. In 1893, before Duddell had done his work, Edward L Nichols had presented a paper entitled 'Phenomena of the time-infinitesimal' to the American Association for the Advancement of Science meeting at Madison, Wisconsin [60]. In the course of this, he referred quite briefly to the work of G S Moler and his use of a galvanometer to record waveforms [61]. Blondel was writing his short article on oscillograph correction entitled 'Remarks on the oscillographic method' [62] to the journal *La Lumière Électrique* at that time and in the introduction to that article he complained that Nichols seemed to be attributing to Moler 'the paternity of the type of instruments called oscillographs', although this word had not been mentioned by Nichols at all. Blondel asserted that he wanted to state once and for all that the word was used for the first time in his own description of this apparatus. In a footnote to another paper in 1905 [63] he referred to himself as the 'father of oscillographs' and claimed that his precedence had been recognised by Duddell himself—as indeed it had in so far as the original idea was concerned. There even seems to have been a legal dispute, or some other such action, as he refers to the fact that 'the Jury of the Exhibition of St Louis (1904) had decided decisively in my favour the question of priority'. In the 'Electrician' letter referred to above, he also quotes some correspondence on the subject between himself and some eminent patent lawyers in Washington.

Be all that as it may, the special correspondent of the *Electrician* at the St Louis Exhibition reported that 'the Duddell oscillograph worked admirably—no small credit to the British Commissioner as it is the kind of instrument often apt to be very unresponsive to anyone but an expert' [64]. It is also worth noting that Duddell was awarded a Gold Medal at the exhibition as a measure of the appreciation of his instrument by the international community of engineers [65].

4.4 Bifilar instruments constructed by other experimenters

Duddell, in partnership with the Cambridge Scientific Instrument Co., clearly made a great commercial success of the bifilar oscillograph, but it must not be thought that he was the only one to undertake development of the original idea. F A Laws, for example, at MIT, had purchased one of Duddell's instruments in 1901 and had become irritated by its extreme delicacy and its susceptibility to damage [66]. He set about designing a more robust version, intended primarily for use in his own laboratory classes. His chief concern was to allow for easy renewal of the bifilar element in the event of failure and his solution to the problem can be seen in figure 4.27. By releasing a thumb screw the inner assembly (shown in the diagram) could be drawn up out of the oil bath. The vibrating loop made of flat strip was soldered to the two copper wires leading to the terminals at the top and it was kept in a state of tension by being looped over the coil spring at the bottom. The natural frequency of vibration of the loop was 6000 Hz. The paper in which Laws describes his instrument says nothing about the method of attaching the mirror to the loop, so one must infer that the time of five minutes quoted by him for replacement assumes that a previously prepared loop and mirror assembly was available, ready for immediate use.

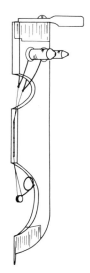

Figure 4.27 Laws' bifilar vibrator designed for easy replacement of the loop in the event of accidental damage.

Two units mounted in the electromagnet are shown in figure 4.28, and figure 4.29 shows the complete apparatus. A rotating mirror was used to throw the trace onto a ground-glass screen. The rotation was produced by turning the handle seen at the front or by coupling it to a synchronous motor. Arrangements were also provided for photographing the trace, the shutter being operated either by hand or by means of a foot pedal.

A R B Wehnelt of Erlangen near Nuremberg, inventor of the Wehnelt

Figure 4.28 Laws' vibrators mounted in their electromagnet [66]. Photograph: Science Museum (391/86).

Figure 4.29 Laws' complete oscillograph. The magnet assembly is at the right: the handle is used to rotate the viewing mirror [66]. Photograph: Science Museum (392/86).

chemical interrupter for induction coils [67,68] considered the Duddell and Blondel oscillographs 'too expensive to come into general use' and in 1903 he constructed the cheap and simple version shown in figure 4.30. Two identical vibrators can be seen mounted on an ordinary laboratory retort stand. The electromagnets producing the field, the pole pieces and the vibrating elements (which were made of hand-drawn silver wire) mounted between the pole pieces can be seen in the photograph. They were used in conjunction with a rotating mirror and the trace was produced on a screen placed about two metres away. With 0.5 A flowing in the loop it was claimed that a deflection of up to 100 cm could be produced on the screen. The loop tensioning system seems to have been fairly primitive and the natural frequency of the vibrating loop was only about 300 Hz. No damping other than some electromagnetic damping seems to have been provided, so it must have been an instrument of very limited application. Wehnelt himself used it to demonstrate the phaseshift effects in inductors and capacitors to his classes. The voltage and current waveforms were displayed simultaneously on the screen and the introduction of inductance or capacitance into the circuit produced a very convincing shift in phase between the two traces.

Another variant is referred to in the literature as the Ganz oscillograph. (The present author has been unable to find a full description of this particular instrument, but it is referred to by several authors writing upon the

Figure 4.30 Inexpensive bifilar oscillograph for laboratory use constructed by Wehnelt [67].

general subject of oscillographs [69,70].) A wire loop similar to that of Blondel was used, but one limb was made to carry the current flowing in the circuit under investigation, whilst the other carried a current proportional to the circuit voltage (by means of a normal 'voltmeter' type of connection with a high series resistance). No mirror was used in this case of course, the images of the wires being formed directly on a screen or on a moving photographic plate. A stroboscopic disc was also used for direct observation, a method which will be described more fully in the next chapter.

Professor Franz Wittmann of the Technical Hochschule in Budapest [71], in a paper written in 1904, described several bifilar instruments which he had built for purposes of the usual waveform display and also for the display of B/H curves of magnetic material using two orthogonal vibrators. D K Morris and J K Catterson-Smith used the Duddell instruments for a similar purpose [72,73].

Several industrial firms also developed their own versions, the so-called 'OSISO' range produced by the Westinghouse Electric and Manufacturing Company being an example [74–76]. These contained very neat permanent magnet vibrators (see figure 4.31(b)) situated horizontally at the bottom of the case of the instrument. The trace could be photographed or viewed with the aid of a very compact optical system including a four-sided mirror drum seen clearly in the view of figure 4.31(a). This oscillograph also contained a rather interesting automatic switching system which enabled one to display different waveforms on the alternate sweeps, persistence of vision being relied upon to produce effective superposition of the two traces. It was suggested by J W Legg in the article referenced above that one of the major uses of this oscillograph was to train deaf-mutes to speak, and enable them to 'receive sound broadcast speeches by sight after sufficient training in this new art'.

Strange to say, the books and paper referenced above pass no comment on the curious name 'OSISO' by which these Westinghouse instruments were known. The secret is revealed in the firm's catalogue, published in June 1925, (Section 3b.),

Its main use is for studying oscillations, hence the first two letters of its name. The illuminant is a straight-filament incandescent-lamp, hence the middle letter of the name. The last 'S' stands for 'scope', the visualising property of this instrument. The final 'O' stands for optical efficiency, a property which is undeniable considering that a few watts from three dry cells, or from the secondary of a bell-ringing transformer, will give results in this instrument which required a 110-volt d.c. arc-lamp in older forms of oscillograph.

It is as well to know: one would never have guessed!

The 'OSISO' appeared in the Westinghouse catalogues for at least five years, but seems to have been dropped sometime in the 1930s. It was a very

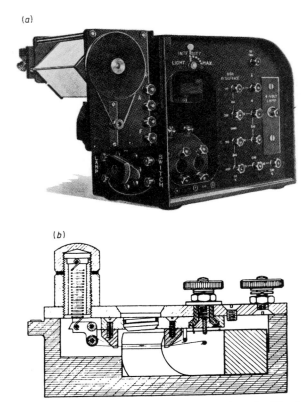

Figure 4.31   The 'OSISO' oscillograph produced by the Westinghouse Co.: (*a*), the complete instrument; (*b*), the bifilar vibrator [74]. Photograph: Science Museum (395/86).

compact instrument, and measured only about 6 in × 9 in × 10 in overall. It was sold at a cost of $750 for the basic model, although optional extras for using various types of recording film were also available.

The Westinghouse Co. also produced other configurations of the oscillograph, one of them, for example, being designed to record fault conditions on electrical mains systems automatically [77]. There were also versions produced by the Siemens Co. [78], and L T Robinson, whose name has been mentioned in a previous chapter, was associated with the design of the version produced by the General Electric Company (USA) in about 1905 [79,80]. This company continued to produce oscillographs for many years after this date [81,82]. Robinson seems to have preferred to use silver for the material of the vibrating strips (see [79]):

The advantage of silver over phosphor bronze which has usually been employed for the vibrating strips is considerable. The resistance, and consequently the energy required to operate the vibrator, is by this means much reduced, which permits the current wave to be taken with shunts having comparatively small fall of pressure (i.e. voltage drop) and the danger of creeping of the reflected image due to expansion of the strips on account of the current passing through them, is entirely overcome. With high-resistance alloys this is one of the limiting features.

Blondel himself seems to have arrived at a similar conclusion. After stating, in one of his papers, that aluminium wire was generally satisfactory, he went on to say 'phosphor-bronze or silicon-bronze also gives satisfactory results and it is very easy to prepare; for greater sensitivity, one still prefers silver.' [83].

4.5 Single-wire instruments

If the bifilar oscillograph of Blondel may be regarded as a simplified form of the moving-coil instrument then an even greater simplification would be to reduce the moving part to a single wire situated in the magnetic field. This is the principle of the 'string galvanometer' which was designed by W Einthoven, Professor of Physiology at Leiden in the Netherlands, to assist his electrocardiographic studies—see figure 4.32 [84–87]. The current under observation passes through a vertical wire situated between the poles

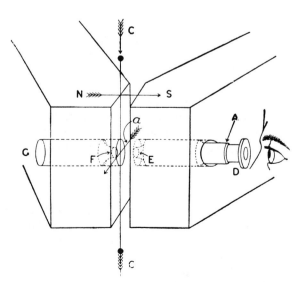

Figure 4.32 Diagram showing principle of operation of the Einthoven string galvanometer [92]. Courtesy: Cambridge Instruments Ltd.

of a strong magnet. For simple DC measurements the front-to-back deflection of the wire can be observed by means of a microscope arrangement fixed into one of the pole pieces. For varying currents it was necessary to project an image or shadow of the wire onto a screen or a photographic plate. A cylindrical Boys' lens could be used in the usual way to compress the vertical shadow into a single spot.

A number of different materials were used for the wire in the various versions of the instrument. Glass or quartz fibre coated with silver or gold was often employed, as were silver and phosphor-bronze. The sensitivity and frequency response depended on the physical parameters of the wire; the length and diameter, the material of which it was made, the tension under which it was held, the strength of the magnetic field etc. The damping employed was simply that resulting from the motion of the wire through the air, together with electromagnetic damping produced by the back EMF, so it was not easy to attain 'dead beat' operation. Indeed, only under certain conditions of circuit resistance could this be achieved. In addition, the motion of the string was not as simple as might be supposed at first sight, there being a tendency to vibrate in several different harmonic modes [88,89].

It was originally intended to be operated at only the very low frequencies encountered in electrocardiographic work but it was further developed so that it could handle rather higher frequencies. J T Irwin [90] quoted some typical figures. For a silver wire, 0.02 mm in diameter, with a resonant frequency of 100 Hz a deflection sensitivity of 1 mm A^{-1} could be obtained, the 1 mm being measured on a screen after an optical magnification of 600 times. It was possible to raise the resonant frequency by tightening the wire but this was accompanied by a decrease in deflection sensitivity. The silver wire was not able to stand much increase in tension, but by substituting phosphor-bronze the resonance could be raised to around 2000 Hz. The Einthoven string galvanometer was produced commercially in Great Britain by the Cambridge Instrument Company, Duddell being largely responsible for its design—see figure 4.33 [91, 92].

The General Radio Company in the USA produced a string oscillograph based on the Einthoven principle [93]. This is shown in figure 4.34. The string assembly itself was situated towards the right-hand end of the long tube. A car headlamp bulb at the extreme right illuminated the string and its shadow image was focused into the box at the left. This contained an eight-sided mirror drum assembly driven by a shaded-pole motor (non-synchronous and therefore variable in speed) to provide the time scan. The image was viewed on a screen at the top of the box. The vibrating string itself was made of tungsten wire of 0.0004 in diameter and 4.5 in long. Provision was made for damping, if required, by the careful insertion of two drops of oil. The string could easily be removed from between the two poles and since there was no mirror mounted on it, it was a simple matter to replace it in the event of damage—much simpler than in most of the

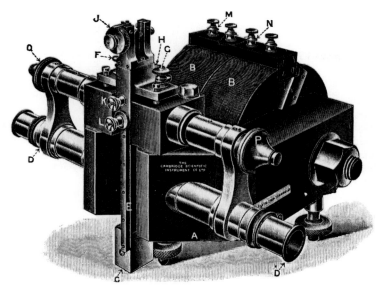

Figure 4.33 The Einthoven string galvanometer as constructed by the Cambridge Scientific Instrument Co. [92]. Courtesy: Cambridge Instruments Ltd.

Figure 4.34 The General Radio Co. (USA) string galvanometer [93].

bifilar loop oscillographs which, as has been mentioned, were difficult and time-consuming to repair. The string was tunable over a wide range, and the following figures are quoted: for a resonant frequency of 60 Hz, 0.2 mV was required to produce an amplitude on the screen of 1 mm. At 1000 Hz, 8.5 mV was required.

References

[1]   Gerard E 1899 *Leçons sur l'Électricité* 6th edn (Paris: Gauthier-Villars) p325

[2]   Shida R 1886 *Phil. Mag.* (Series 5) **22** 96–104

[3]   Moler G S 1892 *Trans. Am. IEE* **9** 223–7 and 227–8 (Discussion)

[4]   Spencer C J 1904 *Electrical World and Engineer* **43** 169–170

[5]   Becker H, Larmayer G and Picard D 1889 *La Lumière Électrique* **31** 16–23

[6]   Gerard E 1890 *Phil. Mag.* (Series 5) **29** 180–2

[7]   Gerard E 1889 *Electrician* **22** 299

[8]   Laws F A 1938 *Electrical Measurements* (New York: McGraw-Hill) pp14ff

[9]   Cornu A 1887 *C.R. Acad. Sci., Paris* **104** 1463–70

[10] Blondel A 1893 *C.R. Acad. Sci., Paris* **116** 748–52

[11] Blondel A 1902 *L'Éclairage Électrique* **33** 115–25

[12] MacGregor-Morris J T and Mines R 1925 *J.IEE* **63** 1056–1107 (*see* Section II (4))

[13] Blondel A 1901 *Rev. Gén. des Sciences* **12** 612–26

[14] Blondel A 1893 *Electrician* **30** 571–2

[15] Blondel A 1893 *C. R. Acad. Sci., Paris* **116** 502–6

[16] Salomonson J K A W 1912 *Electrician* **69** 357–60

[17] Blondel A 1894 *La Lumière Électrique* **51** 172–5

[18] Blondel A 1900 *Congrès Int. de Physique* (*Paris*) *1900* vol 3 (Paris: Gauthier-Villars) Report No 8, pp264–95

[19] Blondel A *see* [13]

[20] Irwin J T 1925 *Oscillographs* (London: Sir Isaac Pitman) p31

[21] Irwin J T *see* [20] p37

[22] Blondel A 1902 *L'Éclairage Électrique* **31** 41–50 and 161–8

[23] Blondel A *see* [18] p286

[24] Skirl W 1928 *Siemens Handbücher* 6 Band. *'Elektrische Messungen'* (Berlin: Walter de Gruyter) p412

[25] Blondel A *see* [13] p620

[26] Blondel A *see* [15]

[27] Blondel A *see* [18] p286

[28] Blondel A *see* [13] pp614–16

[29] Blondel A *see* [22] p49

[30] Blondel A 1897 *L'Éclairage Électrique* **11** 158–9

[31] Duddell W D and Marchant E W 1899 *J.IEE* **28** 1–107

[32] Duddell W D 1898 *GB Patent Specification* 5449

[33] Lamb G C 1906 *Alternating Currents* (London: Arnold) pp127ff

[34] Ayrton W E and Mather T 1898 *Phil. Mag* (Series 5) **46** 349–79

[35] Report 1895 *Engineering* **60** 563

[36] Duddell W D 1897 *Electrician* **39** 636–8

[37] Barron S L 1950 W. D. Duddell, C.B.E., M.I.E.E., F.R.S. and the Cambridge Instrument Co. Ltd *Cambridge Instrument Co., Monograph No 2* (March 1950) *See* Introduction

[38] Report 1900 *Sci. Am. Suppl.* **49** No 1278, pp20482–20483

[39] Report 1900 *Nature* **63** 142–5
[40] Catalogue 1913 *Cambridge Scientific Instrument Co., List No 118* p15
[41] Catalogue 1903 *Cambridge Scientific Instrument Co., List No 20* pp14–15
[42] Report 1903 *Sci. Am.* **88** 468–9
[43] Report 1904 *Sci. Am. Suppl.* **57** No 1482, pp23748–23750
[44] Perkins F C 1904 *Electrical Review* **44** 7–9
[45] Morris J T 1907 *Electrician* **59** 292–4
[46] Ramsay D A 1906 *Electrician* **57** 884–7
[47] Fleming J A 1901 *Handbook for the Electrical Laboratory and Testing Room* vol 1 (*Electrician* 1901) p406
[48] Report *Nature, see* [39]
[49] Perkins F C *see* [44]
[50] Ramsay D A *see* [46]
[51] Barron S L *see* [37]–postscript
[52] Brown B 1931 *Talking Pictures* (London: Sir Isaac Pitman) pp236ff
[53] Blondel A 1900 *GB Patent Specification* 23 417 (accepted finally 1901)
[54] Ramsay D A *see* [46]
[55] Duddell W D and Marchant E W *see* [31]
[56] Duddell W D *see* [36]
[57] Blondel A *see* [22]
[58] Blondel A 1906 *Electrician* **58** 299–301
[59] Duddell W D 1906 *Electrician* **58** 342–3
[60] Nichols E L 1894 *Proc. Am. Assoc. for Advancement of Science, 42nd meeting (Madison, Wisconsin) August 1893* (Salem: American Association for Advancement of Science) pp57–71
[61] Moler G S *see* [3]
[62] Blondel A *see* [17]
[63] Blondel A 1905 *Bull. de la Soc. Int. des Électriciens* **5** 207–18
[64] Report 1904 *Electrician* **54** 49–51
[65] Report 1904 *Electrician* **54** 126
[66] Laws F A 1905 *Electrical World and Engineer* **45** 839–40
[67] Wehnelt A 1903 *Elektrotech. Z.* **24** 703–4
[68] Wehnelt A 1903 *Electrical Engineer* **32** 542–3
[69] MacGregor-Morris J T and Mines R *see* [12]
[70] Laws F A 1917 *Electrical Measurements* 1st edn (New York: McGraw-Hill) pp636–8
[71] Wittmann F 1904 *Electrotech. Z.* **41** 885–9
[72] Morris D K and Catterson-Smith J K 1904 *J.IEE* **33** 1019–27
[73] Morris D K and Catterson-Smith J K 1904 *Electrician* **52** 684–5
[74] Legg J W 1927 *The Electric Journal* (sometimes called *Electric Club Journal*) **24** 268–72, 341–6
[75] Smythe W R and Michels W C 1932 *Advanced Electrical Measurements* (New York: Van Nostrand) p134
[76] Michels W C 1941 *Advanced Electrical Measurements* 2nd edn (New York: Van Nostrand) p213
[77] Report 1933 *Instruments* **6** 193
[78] Skirl W *see* [24] pp411ff

[79] Robinson L T 1905 *Trans. Am. IEE* **24** 185–214
[80] Duncan R L and Drew C E 1929 *Radio Telegraphy and Telephony* (New York: Wiley/Chapman and Hall) pp527–30
[81] Report 1930 *Instruments* **3** 713–15
[82] Report 1931 *Instruments* **4** 362–3
[83] Blondel A *see* [13]
[84] Einthoven W 1903 *Ann. Phys.* **12** 1059–71
[85] Einthoven W 1906 *Ann. Phys.* **21** 483–514, 665–700
[86] Einthoven W 1909 *Pflüger's Archiv für Gessamte Physiologie* **130** 287–321
[87] Laws F A *see* [8] pp35ff
[88] Crehore A C 1914 *Phil. Mag* **28** 207–24
[89] Williams H B 1924 *J. Opt. Soc. Am.* **9** 129–74 and 1926 **13** 313–82
[90] Irwin J T *see* [20] ch 2
[91] Barron S L *see* [37]
[92] Catalogue 1908 *Cambridge Scientific Instrument Co.* List No 53
[93] Report 1928 *Instruments* **1** 491–2
[94] Catalogue 1932 *Cambridge Scientific Instrument Co.* List No 118

5

Moving-Iron Oscillographs and Compensated Instruments

Summary

There were two other classes of mechanical oscillographs using vibrating elements which, although never achieving the same degree of popularity as the bifilar moving-coil type, showed considerable promise when first introduced. Commercial models were available, notably in France.

The first class used a small iron bar as the vibrating element, the movement being controlled by the magnitude of a magnetic field which was, itself, dependent on the current under observation.

The second class, instead of reducing the moving-coil element to a bifilar loop as discussed in the last chapter, used a moving coil of more normal proportions, but employed an external electrical circuit to precondition the current waveform under observation, thereby compensating for the imperfections of the instrument.

5.1 Moving-iron oscillographs

At the same time as he was engaged in constructing his first bifilar loop oscillograph, Blondel was also trying out another idea which involved the use of a moving-iron vibrator. The general principle which lay behind this type of instrument is shown in figure 5.1. If a soft-iron needle mounted on a pivot is situated between the poles of a magnet, it rotates to a position where it is aligned with the magnetic field H_1 and itself becomes magnetised. If it is now subjected to a second field H_2 applied at 90° to the first, produced by a pair of coils situated at front and rear, it will align itself with the resultant field and will lie at some other angle θ. Provided that the angular movement is not too large, the deflection will be approximately proportional to the strength of the second field. In fact,

$$\tan \theta = H_2/H_1 \simeq \theta$$

if the angle is small. If the current in the coils is alternating, and if the mass of the needle, the torsion in the suspension etc. are of suitable magnitudes then the needle deflection will follow the variation of the current, and its movement can be monitored in the usual way by means of a light beam and mirror.

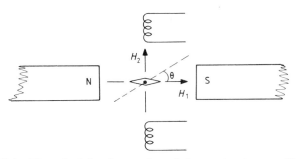

Figure 5.1   The principle of operation of the moving-iron oscillograph

Blondel's first idea seems to have been to use a horseshoe magnet with a small iron plate fixed to one end of one of the pole pieces, being free to vibrate past the other pole piece as shown in figure 5.2. This he described as 'a developed form of telephone receiver' [1,2]. However, this does not seem to have been an especially successful arrangement and little more was heard of it after an initial description in 1893. It was soon replaced by a tiny bar of soft iron mounted on pivots between the pole pieces [3–6]. The use of a spring mounting or a bifilar form of suspension was also mentioned as a possibility. The plan of his first instrument of this type is shown in figure 5.3. Two pole-piece extensions P concentrated the main field into a small gap some 2 or 3 mm wide and the transverse field was produced by the two coils B and B'. In order to reduce any eddy currents when used with AC the pole-piece extensions were laminated horizontally. The moving soft-iron bar was made tall and thin in shape in order to minimise the inertia

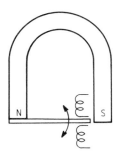

Figure 5.2   Blondel's first suggested form of moving-iron oscillograph described as 'a developed form of telephone receiver'.

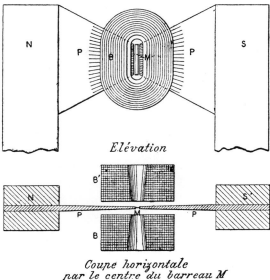

Elévation

Coupe horizontale
par le centre du barreau M

Schéma de l'oscillographe à fer doux de 1893.

N, S, pôles d'un aimant ou électro-aimant. — P, pièces polaires feuilletées. —
B, B', bobines à courants alternatifs. — M, barreau de fer doux oscillant,
portant un miroir.

Figure 5.3 Blondel's   first   practical   moving-iron   oscillograph
vibrator—1893 [1].

of the system. As Blondel explained, this would be more sensitive than a
single needle, consisting in effect of many needles stacked one above the
other. An oscillograph of this type was constructed for Blondel by the Paris
firm of Pellin and was exhibited at the exhibition which was held in that city
in 1897. The complete instrument can be seen in figure 5.4; the layout and
optical system are shown in the plan view of figure 5.5. This was a double,
or two-channel oscillograph, the two moving-iron vibrators being situated
at O_1 and O_2. Light from the filament lamp F is split by the prism arrangement, one half of the beam being cast upon each vibrating element which,
of course, has a tiny mirror mounted on it. The reflected beams return to
the second set of prisms and are brought to a focus upon the ground-glass
screen V situated at the end of the tube L, the exact position of which can
be varied by a rack and pinion mechanism in order to adjust the focus
exactly. A slotted stroboscopic disc is used here to provide the time scan in
place of the more usual rotating or vibrating mirror [7,8]. This is illustrated more fully in the schematic diagram of figure 5.6. A vertical slit
illuminated by the lamp is scanned by the slots in the disc and the movement
of the slots provides a vertical scanning of the spot across the ground-glass

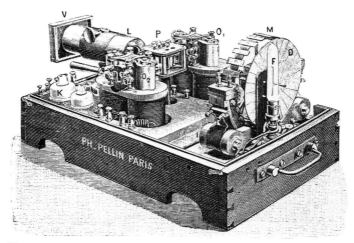

Fig. 7. — *Grand oscillographe double à fer doux, modèle de 1897.* — F, lampes
à incandescence; P, disque tournant entraîné par le moteur synchrone
M; O_1, O_2, oscillographes à électro-aimants; P, prismes pour envoyer les
rayons lumineux incidents sur les miroirs des deux galvanomètres, et ren-
voyer les rayons réfléchis dans le tube L; V, écran d'observation des courbes.

Figure 5.4 Two-channel Blondel moving-iron instrument—1897 [3].
Courtesy: Gauthier-Villars, Paris.

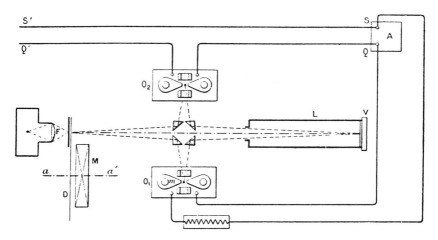

Oscillographe double de 1897. — Vue d'ensemble et de montage.

A, appareil à étudier. — S, Q, conducteurs du courant. — F, appareil optique,
projecteur et fente verticale (ou lampe à incandescence). — D, disque à fentes
radiales. — M, moteur synchrone. — O_1, O_2, oscillographes (voltmètre et ampère-
mètre). — L, tube noirci. — V, écran ou plaque photographique.

Figure 5.5 Optical and electrical arrangements of the moving-iron
oscillograph [3]. Courtesy: Gauthier-Villars, Paris.

screen. (N.B. Figure 5.6 shows the optical path without the two 90° turns introduced by the prisms in the actual apparatus.) If the disc is large and the angular span on the slit is not too great a reasonably linear scan is produced although, as Blondel remarked, 'this method only produces a fairly feeble luminous spot and does not allow extreme reduction in the size of the oscillograph mirrors'. The slotted disc, marked D in figure 5.4, is driven by a synchronous motor M of open construction, the teeth of which can be seen arranged around the periphery of the rotor. As in most of the oscillographs described in the previous chapter, the trace on the screen has a vertical time axis—i.e. is rotated by 90° from that considered normal today. A photographic plate could be inserted in place of the ground-glass screen if required.

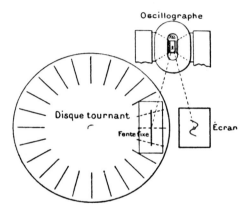

Figure 5.6 Slotted-disc method of producing time scan [3]. Courtesy: Gauthier-Villars, Paris.

This instrument of Blondel's, dating in conception from the early 1890s seems to have been anticipated by R Colley [9] who published details of a system working on similar principles in 1885,—see figure 5.7. A magnetised steel plate p was suspended on a vertical silk thread in the centre of a solenoid G. The precise details of its operation are not totally clear from Colley's paper, but it seems to have been a tangent galvanometer type of instrument using the earth's magnetic field as its restoring force. When an oscillating current passed through the coil, it was set into vibration and these movements were observed by means of a mirror mounted on the steel disc. Sunlight was allowed to fall upon the large disc in front of the apparatus, this disc being caused to rotate by means of a clockwork motor. A small hole pierced in the disc allowed a shaft of light to fall upon the mirror, the movement of the disc producing a time scan. The movement of the mirror combined with that of the spot produced a wave trace when viewed through the microscope M. A micrometer could be used with the microscope to allow measurement of the wavelength of the trace. This apparatus was used to investigate transient oscillations in *LC* circuits and, to aid in this, a triggering apparatus was situated in front of the disc. A platinum point dipped down through a small glass pot of alcohol to rest on the metal bottom of the pot. A small projection (a) on the rim of the disc struck the lever on which the point was mounted once in every revolution, breaking the circuit and creating a transient which commenced at the same point in each revolution so that a synchronised trace would appear.

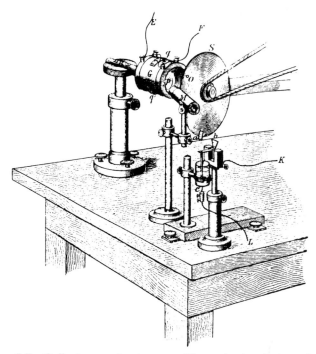

Figure 5.7 Colley's moving-iron oscillograph (predating that of Blondel) [9].

Returning now to Blondel's oscillograph; the vibrating element itself can be seen in more detail in figure 5.8. The oscillating bar was 1 mm wide and 15 mm high and was mounted on fine pivots between two bearings, the precise setting of which could be adjusted by screw threads. The whole thing was contained in an ivory or German silver (a high resistance copper–zinc–nickel alloy) tube filled with oil for damping and provided with a lens to permit the passage of the light beam to and from the mirror mounted on the bar. A relatively large mirror having an area of 15 mm^2 was used in the interests of adequate illumination; this increased the inertia and hence reduced the resonant frequency of the vibrating element. The earliest model, constructed in 1892, had a resonant frequency of only 1500 Hz, but by the time of the 1897 exhibition this had been increased to 5000–6000 Hz. Being freely mounted on pivots there was no mechanical restraining torque, apart from the restoring force of the field of the main magnet which, in this case, was an electromagnet. The curve of figure 5.9(*a*) shows the resonant frequency as a function of the electromagnet current. With increasing current the resonant frequency rises until the point where saturation occurs in the vibrating bar, when it attains a value of 5500 Hz. (N.B. 'double vibrations' are plotted here, meaning complete cycles.)

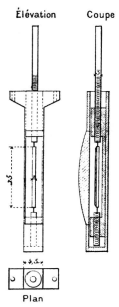

Élévation   Coupe

Plan

Figure 5.8   The vibrator of the Blondel oscillograph [3].
Courtesy: Gauthier-Villars, Paris.

In the very earliest models, Blondel tried to produce the required damping by the use of rubber buffers. This was not a success and he then went on to use various fluids, including water, glycerine and various balsams. He seems to have settled eventually on either vaseline oil or castor oil (huile de ricin). The degrees of damping which could be obtained with the two types of oil under various conditions of magnetic field strength are shown in figures 5.9(*b*) and (*c*). These show the responses to a 50 Hz square-wave current.

This instrument was quite adequate for the observation of waveforms of 'industrial frequencies' and, moreover, was able to show two curves superimposed simultaneously and in the correct phase relationship, the first

(a)

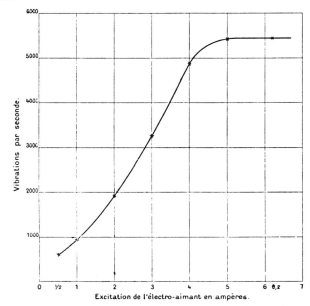

Nombre de vibrations doubles, en fonction de l'excitation
de l'électro-aimant.

(b)

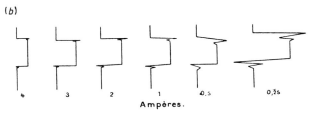

Courbes d'amortissement du barreau dans l'huile de vaseline,
sous des excitations variées.

(c)

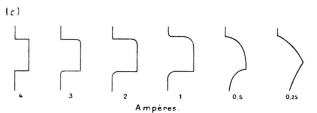

Courbes d'amortissement dans l'huile de ricin,
sous des vibrations variées.

Figure 5.9 (*a*) Resonant frequency of the moving-iron vibrator,
shown as a function of the electromagnet current. (*b*) and (*c*) Step
responses of the instrument with vaseline oil and castor oil damping
respectively, for various values of electromagnet current [3]. Courtesy:
Gauthier-Villars, Paris.

time that this had been achieved. The circuit diagram of figure 5.5 shows an alternating supply voltage S′Q′ connected to a piece of apparatus A, the upper oscillograph element being connected as an ammeter, the lower, via a series resistance, being connected as a voltmeter. The waveforms of figure 5.10 show the voltage and current of an arc lamp obtained in this way.

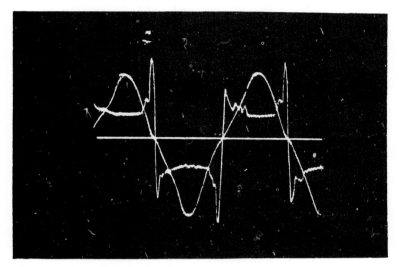

Figure 5.10   Voltage and current waveforms of an AC arc in an inductive circuit, displayed by the moving-iron oscillograph [3]. Courtesy: Gauthier-Villars, Paris.

In an attempt to increase still further the resonant frequency, Blondel considered replacing the pivots by a vertical torsion wire. These pivots had obviously caused him a great deal of trouble; as he stated, 'In spite of the great progress made, the moving-bar oscillographs still gave frequencies which were too low for my taste, and the use of pivots, in spite of their perfection, were subject to some inconveniences.' [10]. One can indeed imagine the difficulty of providing pivots which were sufficiently small to be used with a bar which was only 1 mm wide, yet would withstand the considerable magnetic forces generated. In considering the possibility of mounting the bar upon a torsional band or strip, he hit upon the idea of making the strip itself the vibrating element [11–19]. In this way he was able to increase the resonant frequency to no less than 40 000–50 000 Hz.

Figure 5.11 shows the construction of the active element of this vibrating-strip model and figure 5.12 shows two such elements mounted up with pole pieces and various adjustment screws. The outer tube (T in figure 5.11) is made of ivory or German silver. Two pole pieces P are set into this tube so that the magnetic air gap is as small as possible and a small window is provided for viewing the mirror m. The toothed wheels D engage with the

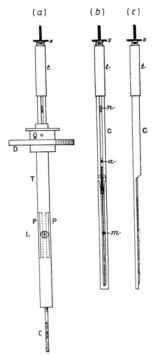

Fig. 10. — *Boîte à huile du nouvel oscillographe à bande de fer vibrante.* — T, tube à huile en ivoire; P, P, pièces de concentration du champ magnétique, incrustées dans l'ivoire (ou fixées au support); en avant se trouve une fenêtre munie d'une lentille l., également incrustée dans la paroi; Q, anneau de cuivre muni d'une vis tangente D; C, détail du chevalet-support en ivoire contenu dans la boîte à huile; *m*, miroir collé sur la bande; *a*, attache de la bande à un petit chariot qui coulisse dans la rainure rectangulaire C; *n*, tige d'attache de ce chariot; *t*, petit cylindre en cuivre, contenant un ressort spiral autour de la tige; *s*, bouton moleté servant d'écrou à la partie supérieure filetée de la tige *n*; en tournant ce bouton on tend plus ou moins le ressort qui soulève la tige, et, par suite, on tend plus ou moins la bande de fer doux fixée en *a*.

Figure 5.11 The vibrator of the iron-band oscillograph. Part (*a*) shows the inner tube C with band inserted in the outer tube. Parts (*b*) and (*c*) show front and side views, respectively, of the inner tube withdrawn from the outer tube [5].

worm wheels V of figure 5.12 and provide zero adjustments for the traces. Vertical alignment is achieved by screws v at the base of the assembly. The small mirror M, the angular position of which is adjusted by R, is used to draw the reference time axis on the final display. Returning to figure 5.11; the vibrating band itself which is 2 or 3 mm wide is contained in a rectangular channel cut into the inner ivory tube C. The upper end is attached to a slider (a) which in turn is fixed to a shaft lying in the channel. The copper tube t at the top contains a spiral spring, the lower end of which is fastened to the shaft at n. Adjustment of the nut s alters the tension of the spring and hence the resonant frequency. The inner tube C slides into the outer tube T which also contains damping oil. The tiny mirror fixed to the band is only 0.2 mm wide by 0.5 mm high by approximately 0.1 mm in thickness, being stuck on with shellac.

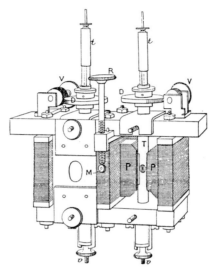

Fig. 12. — *Détail de la figure 11. Vue des pièces rapportées entre les pôles de l'aimant.* — P, P, pièces polaires feuilletées en tôle; T, tubes à huile contenant les équipages mobiles; v, v, vis-écrous moletés pour élever ou abaisser les tubes; V, vis sans fin attaquant les roues D pour orienter les tubes; M, miroir destiné à donner le trait de repère; R, bouton pour orienter ce miroir.

Figure 5.12 Two iron-band elements mounted with pole pieces, and showing the various adjusting screws [5].

Figure 5.13 shows the complete double-vibrator assembly fixed in position on the magnet which produces the main field. One of the deflecting coils has been removed from the right-hand element and this can be seen (B) lying alongside. The shaped piece of soft-iron S is a magnetic shunt which can be placed across the poles of the magnet if required, thereby reducing the resonant frequency but giving increased sensitivity. As stated above, with an electromagnet providing a concentrated field for a single vibrating element, a resonant frequency of over 40 000 Hz could be attained. In this double-vibrator instrument the field was rather less intense and a frequency up to 25 000 Hz was possible. The curves of figure 5.14 show the variation of resonant frequency and sensitivity with strength of the main magnetic field for this type of iron-band vibrator. The units of the magnetic field axis are in amperes (through the coil of the electromagnet) and the sensitivity (left-hand scale) is given in millimetres deviation of the

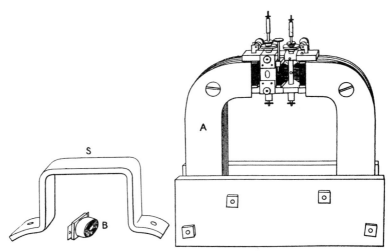

Fig. 11. — *Vue générale de l'oscillographe double à bande de fer.* — A, aimant carré ; S, barre de fer qu'on peut placer à cheval sur les pôles de l'aimant pour former shunt magnétique, et réduire le champ magnétique et le nombre des vibrations propres ; B, une des bobines, détachée du galvanomètre de droite.

Figure 5.13   The complete iron-band oscillograph assembly. The coil B has been removed from the front of the vibrator on the right [5].

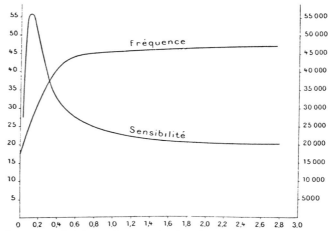

Fig. 13. — *Spécimen d'une courbe de relation entre les intensités du champ magnétique, la sensibilité et la fréquence.* — Abscisses : courants d'excitation du champ (électro-aimant) en ampères ; ordonnées de droite : nombres de vibrations doubles par seconde ; ordonnées de gauche : déviations à 1 mètre en millimètres par ampère (avec de très petites bobines de 6 ohms de résistance).

Figure 5.14   Curves showing the sensitivity (left-hand scale) and resonant frequency (right-hand scale) as functions of the electromagnet current [5].

light beam at a distance of 1 m. The frequencies (right-hand scale) are in cycles/second (Hz).

> To avoid the complications of setting it up for each different experiment, and to make the whole thing portable, all the various items were fixed inside a solid wooden case. This case, which is given the name "Kodak" by analogy with the photographic apparatus of Eastman, forms a dark-room.

The portable arrangement described thus by Blondel [20] can be seen in figures 5.15 and 5.16. A slightly different version constructed by Carpentier of Paris was shown at the Paris Electrical Congress Exhibition in 1900. The optical arrangement is quite conventional, with the lamp L housed in an external casing (compare with figure 4.8) illuminating the double oscillograph O. The reflected beams passed through a cylindrical lens and onto a rocking mirror and occluding shutter driven by a synchronous motor which reflected it upwards to the hatch D where various accessories for tracing or for photography could be inserted. It was also possible to use a mirror set at 45° to project the trace onto a larger screen for group viewing. These various optical arrangements can be seen in figure 5.17. When used

Fig. 15. — *Vue d'ensemble extérieure de l'oscillographe « Kodak » double.* — C, couvercle de la caisse, auquel est fixée la chambre noire à soufflet P, dont la queue peut se rabattre dans le plan du couvercle; L, lampe à arc; r, ouverture pour la mise en marche du synchronoscope; T, tableau de distribution, contenu dans un placard; V, voltmètre; A, ampère-mètre; B, bornes, et J, commutateur du courant principal, passant par l'ampèregraphe; C, commutateur pour mise en court-circuit de l'ampère-mètre; b_1, bornes du circuit voltgraphe; b_2, bornes du circuit du synchronoscope; F, fusibles du circuit voltgraphe et du synchronoscope; m, n, p, interrupteurs des circuits du voltgraphe, du voltmètre et du synchronoscope.

Figure 5.15 Complete exterior view of the Blondel 'Kodak' double moving-band oscillograph [5].

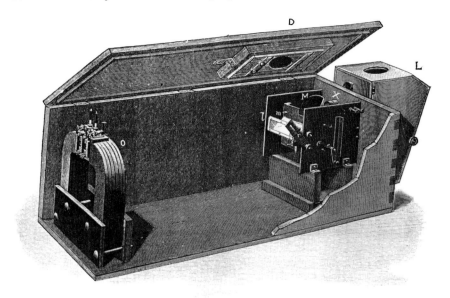

Fig. 16. — *Vue intérieure du même appareil « Kodak » double.* — O, aimant à lames, portant à sa partie supérieure les deux oscillographes jumeaux ; D, chambre noire à soufflet ; *l*, lentille cylindrique ; *m*, miroir oscillant du synchronoscope ; M, disque obturateur monté sur l'arbre et tournant devant l'objectif ; X, objectif à lentille cylindrique fixé dans la paroi de la caisse et muni d'un diaphragme à trois fentes éclairant respectivement les miroirs des deux équipages mobiles et le miroir de repère.

Figure 5.16 Interior view of the oscillograph of figure 5.15 showing the moving-band elements on the left and the oscillating mirror 'synchroscope' on the right [5].

for projection, it was recommended that a cylindrical lens of longer focal length should be employed to ensure sharp focus on the distant screen, and also that vibrators with rather larger mirrors should be used to provide adequate brightness. An incandescent lamp could be used as a light source, and an acetylene flame was also mentioned as a possibility but the arc lamp was essential when used for projection.

The iron-band oscillograph was capable of high-frequency performance but was not as sensitive as the bifilar loop instrument. It also had the disadvantage that the front and back coils which carried the current under observation possessed inductance which could confuse the situation somewhat. Blondel pointed out that the inductance, in effect, increased the damping term in the equation (4.2), altering it from

$$K\frac{\mathrm{d}\theta(t)}{\mathrm{d}t} \qquad \text{to} \qquad \left(K+\frac{TL}{R+R'}\right)\frac{\mathrm{d}\theta(t)}{\mathrm{d}t}$$

where L represents the inductance of the coils, R is the resistance of the iron band itself and R' is the resistance of the rest of the circuit in series with the band. Thus the effect of the inductance could be counteracted by the

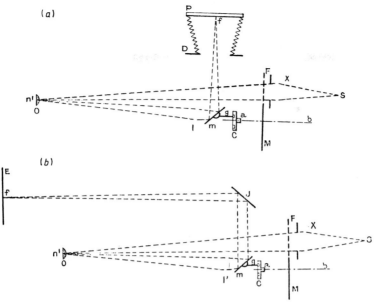

Fig. 17 et 18. — *Schémas de la marche des rayons lumineux dans l'oscillographe « Kodak », soit pour le tracé et la photographie des courbes (fig. 17), soit pour leur projection dans un cours (fig. 18*. — S, source de lumière (arc électrique); **X,** objectif ou condenseur, à lentille cylindrique horizontale; F. diaphragme percé de fentes verticales éclairant chaque petit miroir *n ;* N, miroir plan d'un équipage mobile; O, petite lentille plan-convexe de la boite à huile; *l*, lentille cylindrique horizontale pour la concentration des rayons réfléchis ; *m*, miroir oscillant à axe horizontal (perpendiculaire au tableau) commandé par un levier *g ; ab*, arbre moteur; C, came calée sur cet arbre et agissant sur l'extrémité du levier *g*, et calculée de façon à ce que le déplacement du point lumineux f sur l'écran P soit proportionnel au temps ; M, disque calé sur l'arbre *ab*, et échancré de façon à n'obturer les rayons que pendant le retour du point *f; J,* miroir qui remplace la chambre noire pour les projections; *l'*, lentille cylindrique à long foyer ; E, écran de projection.

Figure 5.17   The optical system of the instrument of figures 5.15 and 5.16: (*a*), arranged for photographic recording; (*b*), for screen projection [5].

choice of a lower value for K—i.e. by using an oil of lower viscosity. Alternatively, if the particular oil in use happened to have too low a viscosity, the inductance could prove to be a positive advantage in increasing the damping to that value required for the critical 'dead beat' condition. Another way of compensating for the inductance was to connect a capacitor across the external series resistor, thus altering the damping term to

$$\left(K + \frac{TL}{R+R'} - GCR'\right)\frac{d\theta}{dt}.$$

Correct selection of C could cancel out the inductive term or could be used to adjust the overall damping to the state required [21,22]. Blondel does not elaborate on this idea to any great extent in his various published papers, and one senses that perhaps he thought that the need to adjust the parameters in this way was a bit of an embarrassment, adding an air of uncertainty in the mind of the user and undermining confidence in the traces produced. In what was possibly an attempt to allay the fears of the users of his instrument, he pointed out that 'the damping and inductance, within their usual range of values, normally provide sufficient precision to render the use of a compensating capacitor unnecessary'. He added that the self-capacitance of the coils would normally be enough for this purpose. This was obviously another complicating factor which reduced the certainty of the operation. In another account he said 'self induction need only be feared if one employs the oscillograph as a voltmeter and uses fine wire coils (i.e. coils with a large number of turns to provide adequate sensitivity) but as a rule it is possible to reduce the drag caused by self induction to a small amount by the addition of a tolerably large dead resistance'. We shall be returning to the whole question of external compensation later in this chapter.

As in the case of the bifilar instrument, it was recommended that the vibrating-bar oscillograph should only be used up to frequencies of $\frac{1}{30}$th of the resonant frequency ($\frac{1}{50}$th in another account).

Blondel summed up the properties of his two classes of oscillograph in the following way [23]:

> the iron-band type is robust and convenient, especially for industrial purposes. Permanent magnets can be employed, hence obviating the need for auxiliary d.c. current. Thanks to the high frequencies which can be achieved (up to 50,000 periods per second) and the compensation of the inductance of the coils it is useful also for laboratory work where one is not dealing with very small currents (below 1/10th amp). The bifilar type, more delicate in construction and less rapid in its oscillations, needs a powerful electromagnet and seems applicable only to laboratory work; it has the advantage of negligible self-inductance and above all has high sensitivity which enables one to use it for the study of currents of less than 1/10th Ampere.

It is worth noting that vibrating-band instruments of this type were still being manufactured until very recently, and the present author has a multi-channel ultraviolet recorder still in use in his laboratory which uses replaceable elements very similar in appearance to those of figure 5.11.

Another French engineer, Monsieur R Dubois, took a critical look at the Blondel moving-iron oscillographs in the early 1920s and, as a result, he was able to improve their design considerably. He pointed out that their relative insensitivity was due mainly to the fact that the coils carrying the current under observation were creating their deflecting field in a magnetic circuit consisting largely of air. He therefore redesigned the instrument, replacing the iron band by a thin iron blade (of area approximately 2 cm^2) balanced on knife edges at the centre of a solenoid carrying the current under observation. In one version the steady restoring cross field was provided by a permanent magnet; in another, shown at the French Exhibition of Physics and Wireless in 1923, by an electromagnet of quite modest proportions. The deflecting solenoid was surrounded by iron, and the presence of the blade within it meant that the air gap was reduced to only about 1 mm. The blade was held in place on its pivots by means of restraining springs. Its movements were observed by means of a pivoted mirror situated well outside the solenoid, connected to one end of the blade by means of a taut wire wrapped around the mirror axle. This resulted in a very neat and compact instrument. It was constructed by the firm of Beaudouin of Paris (Mécanique de précision) and was, presumably, commercially available for those who wished to purchase one.

In the paper which he wrote in volume 17 of the journal *Revue Générale de l'Électricité* in 1925 (pp 977–86), Dubois claimed that his

> oscillographe à palette ... attains 50 times the sensitivity of the best bifilar oscillographs containing a very powerful electromagnet, although it uses only a small permanent magnet. This sensitivity ... is due largely to the use of a large fixed coil, the flux of which is channelled through the oscillating blade. It is equally the consequence of the mechanical amplification of the displacements as they are transmitted to the mirror.

It achieved a natural frequency of oscillation of about 3000 Hz, and the comparison which Dubois made between it and one of the Blondel instruments is very revealing, for he claimed that the Blondel model had a natural frequency of 5000 Hz. This, it will be noted, is much less than the rather elevated frequencies quoted on the graph of figure 5.14, and one therefore suspects that the results of that figure pertain to a special 'one-off' laboratory instrument. Also, Blondel always quoted f_0 in air, and studiously ignored the considerable reduction due to the damping fluid.

Other experimenters constructed their own versions of the moving-needle or moving-bar type of oscillograph. For example, H J Hotchkiss and F E Millis in 1895 made an instrument having a 1 mm wide and 2 cm high bar made of soft iron suspended between pole pieces on a quartz torsion fibre [24–28]. By comparing the natural vibration with a tuning fork they established that the resonant frequency of their vibrating system was 3580 Hz. Later versions went to higher frequencies but, as Blondel pointed out, their instruments made no provision whatsoever for damping and thus they could not provide the ideal operating conditions and ensure traces of high accuracy. Hotchkiss and Millis themselves claimed, however, that when studying waveforms of fundamental frequency 120 Hz 'it was not possible to detect the slightest distortion of the curve due to the combination of the period of the needle with that of the current'—presumably by comparison with some other method of determination, although they do not say this explicitly in their paper.

A similar type of oscillograph was constructed by F J A McKittrick [29], but this only attained a natural frequency of 2656 Hz. Like Hotchkiss and Millis, he used it to study transient phenomena in AC circuits. Another variant on the theme of the vibrating bar was the oscillograph produced by the Westinghouse Electric and Manufacturing Company illustrated in figure 5.18 [30,31]. This was constructed on the 'balanced-armature' principle [32]. The magnet was provided with divided pole pieces and the vibrating bar itself was held in the position shown by means of a torsion fibre (the fibre being in a direction perpendicular to the paper). When a current passed through the coils, one of the pole projections was strengthened and the bar rotated. Electromagnetic damping was used, the degree of damping being adjustable by means of a resistance slider. As far as can be determined, this appears to have been developed for some specific application only, and the present author has been unable to discover the existence of any commercial oscillograph which used the balanced element.

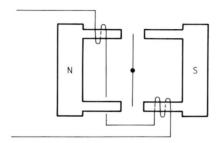

Figure 5.18   The Westinghouse balanced-armature system.

A particularly interesting form of moving-iron instrument was described by R Goldschmidt in 1905—see figure 5.19 [33,34]. If a small pointed needle is placed between the poles of a powerful magnet with its point touching the centre of the face of one of the poles it will stand vertically on the pole face. If it is then displaced slightly from the vertical as in the second diagram the magnetic forces will restore it to the vertical position as soon as it is released. A displacement can also be produced magnetically by means of two coils C_1 and C_2 situated on either side of the needle as in the third diagram and, subject to the usual limits set by mechanical inertia, the displacement of the tip of the needle will follow the waveform of the current in the coils. If the top of the steel needle is polished its movement can be followed by means of a light beam. The practical instrument constructed by Goldschmidt (figure 5.20) was furnished with two pairs of deflecting coils $90°$ apart, one energised from the AC supply through the resistor R and the other through the choke coil T. As well as producing a phaselag, the choke had the effect of filtering out any harmonics in the

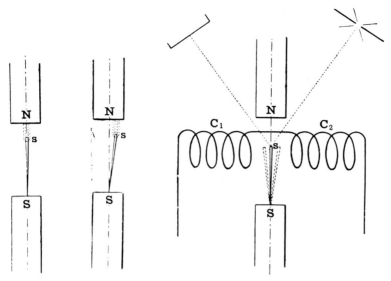

Figure 5.19 Illustrating the principle of operation of Goldschmidt's 'oscillographic wave tracer' [33].

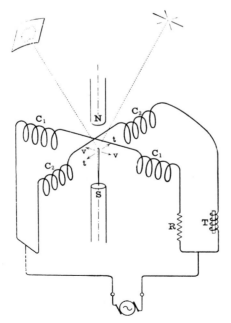

Figure 5.20 Goldschmidt's oscillograph used for the display of 'cyclograms' [33].

waveform and ensuring that the coils C_2 were supplied with a reasonably pure sinusoidal current. The result was that the movement of the needle traced out a Lissajous' figure on the screen. This 'cyclogram' could simply have been used in that form for purposes of measurement, but a rather subtle modification to this simple arrangement was suggested; namely replacing the flat screen by one which was semicircular in the direction of the C_2 deflection as shown in figure 5.21. The sinusoidal motion of the light beam is then seen as a linear motion with respect to time—i.e. the scan has been made to provide a linear time base. No flyback arrangement was incorporated; the trace scanned first in one direction and then in the other producing two superimposed traces, one reading left to right and the other vice versa. It was also suggested that two iron pins could be affixed to the upper pole of the magnet, thereby distorting the magnetic field and linearising the scan. Details are not given however, and one infers that this did not prove to be a very workable system in practice. It was claimed that the necessary damping could be provided electromagnetically by the eddy currents induced in the N pole face by the movement of the needle and, further, that the

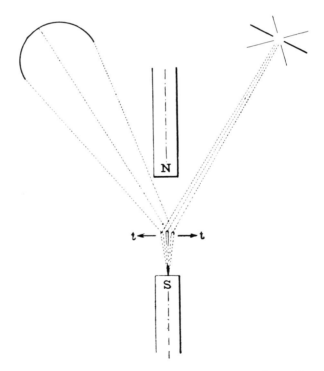

Figure 5.21   The use of a semicircular screen to produce a linear time scan [33].

degree of damping could be adjusted by means of a small copper plate mounted on the pole.

Goldschmidt seemed to place considerable importance on getting a phaseshift of exactly 90° between the two sets of coils, and to this end a rather more elaborate phaseshift circuit with two chokes was proposed. It is not easy to see why an exact 90° shift was so important; perhaps he wanted to display one complete half-cycle precisely inverted over the half-cycle of the opposite polarity for purposes of comparison. Measurements would still have been possible even if they were not precisely superimposed, but the figure might have been a bit more complicated to interpret.

Figure 5.22 Goldschmidt's oscillograph. Note that the vibrating needle (not visible in this view) is horizontal: the large outer coils at the front and back produce the steady DC magnetising field [33].

The apparatus itself can be seen in figure 5.22. The large outer coils produced a steady magnetic field; the deflecting coils produced vertical and horizontal fields. The vibrating needle was $\frac{1}{8}$ in long and its natural frequency of vibration was 5000 Hz, making it quite suitable for viewing waveforms of 50 or 60 Hz. The Editor of the issue of the *Electrician* in which the description of the apparatus appeared made the obvious comment [35] — 'It is a pity that Mr Goldschmidt has omitted a reference to the self-induction of the instrument, as judging by the design this item might not be negligible'. According to that account, the instrument was manufactured by Messrs Crompton and Co. and it could be supplied with two needles and two sets of deflection coils for the simultaneous observation of current and voltage.

Mention was made earlier of the use of electromagnetic damping and of the possibility of adjusting the damping electrically by means of external resistors and capacitors. It is perhaps appropriate at this stage to look at the whole question of compensating networks and, in particular, at some instruments introduced by Professor Henri Abraham which were known as 'rheographs', from the Greek rheo—a current and graphos—to draw.

5.2   Externally compensated instruments

The importance of adjusting the damping of the instrument to the optimum value if an accurate inscription of the waveform was to be produced was stressed by Blondel and, as previously mentioned, one of Duddell's major contributions was the introduction of the flat bifilar strip which presented a greater cross sectional area to the damping fluid. The use of an oily damping fluid in this way was clearly a messy and inconvenient business. The moving element had to be enclosed in a sealed leak-proof chamber and a window had to be provided to permit optical observation of the moving mirror. The presence of several optical interfaces, combined with the dispersion of the light beam in its passage through the oil, reduced the efficiency of the optical system. The damping effect was also critically dependent on temperature and it will be recalled that elaborate warm-up procedures were adopted to ensure the correct viscosity. The presence of the damping oil also tended to reduce the natural resonant frequency of the element and introduced phaseshifts between the applied signal and the recorded trace. Over and above all these practical considerations, the assumption that the damping effect was proportional to $d\theta(t)/dt$ was, at best, a simplification of a rather complex situation. This being so, it seemed to be rather undesirable to place so much reliance on the attainment of a precise value for this parameter.

It was pointed out earlier that when a coil or bifilar loop is in motion in a magnetic field a back EMF is produced and the resulting current acts as a brake or damper to the motion. In the oscillograph instruments this was usually considered to be very small and its effect could be included with the $d\theta(t)/dt$ term in the differential equation. Some control of damping could be provided by changing the resistance and hence altering the magnitude of the reverse braking current. If the instrument were overdamped, the insertion of resistance would tend to correct it; if underdamped a parallel shunt resistance was required [36]. This was not really a very practical method of control since in oscillograph elements the effect was very small and the use of series and parallel resistors naturally affected the overall sensitivity of the system.

Some improvement can be obtained by the use of a resonant parallel circuit, a method described by J T Irwin in his book *Oscillographs* published in 1925 [37]. An *LCR* circuit is connected across the oscillograph loop as shown in figure 5.23. The conditions for optimum damping were derived by Irwin and his colleague Mr Hodgson. The details of the somewhat complicated mathematical analysis need not detain us here, but one of the most important conclusions was that the resonant frequency of the shunt circuit should be the same as the natural resonant frequency of the moving element. The action of the shunt can be explained in a general way as follows. The actual current under observation is no longer passing through the coil of the oscillograph itself. The current is being preconditioned by the presence of the shunt in such a way as to compensate for the imperfect response of the instrument.

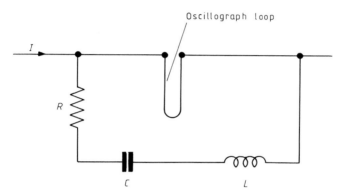

Figure 5.23   Irwin's resonant shunt to achieve optimum conditions of damping [37].

This system seems to have worked quite well when applied to an instrument with a simple vibrating element such as the Einthoven string galvanometer [38]. However, in the bifilar type, the presence of two strings coupled at their centres by the mass of the mirror, the two strings inevitably having slightly different operating parameters, made it impossible to provide proper compensation with this type of shunt [39]. Further analysis revealed that Irwin and Hodgson's analysis had actually neglected the back EMF effect, although this did not invalidate their conclusions in circuits having relatively large resistance. In 1932 E J Martin and D F Caris carried out a more detailed study of bifilar instruments and devised the double-resonant shunt of figure 5.24 in order to overcome these problems [40]. According to their published results, it appeared to work quite successfully and did away with the need for damping oil. However, there seems to be no evidence that their method was widely adopted. One suspects that the two resonant circuits were very fiddling to adjust accurately and, in any case, by that date the cathode ray oscilloscope was beginning to be a serious challenge to the supremacy of the mechanical oscillographs.

The idea of preconditioning was introduced by Henri Abraham in 1897. In order to understand the principle underlying his system it will be necessary to return to the differential equation governing the movement of a moving-coil galvanometer

$$P \frac{d^2\theta(t)}{dt^2} + K \frac{d\theta(t)}{dt} + T\theta(t) = Gi(t). \tag{5.1}$$

It will be recalled that the Blondel/Duddell oscillographs aimed to make the inertia P and the damping K small compared with the other terms so that the equation reduced to the approximate form

$$T\theta(t) = Gi(t) \tag{5.2}$$

thereby ensuring that the deflection $\theta(t)$ followed the applied current $i(t)$ exactly. This led to the design of instruments with small-sized coils and very tight suspensions giving a high natural frequency of vibration.

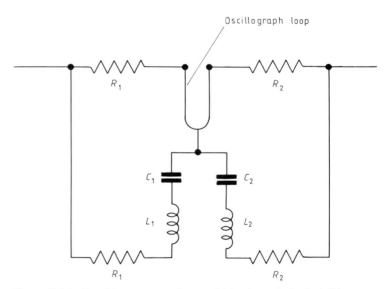

Figure 5.24   Double-resonant shunt of Martin and Caris [40].

Abraham argued rather differently. His approach was to let the inertia term P and the galvanometer sensitivity G dominate the others so that the equation was reduced approximately to

$$P \frac{d^2\theta(t)}{dt^2} = Gi(t) \tag{5.3}$$

The current waveform $I(t)$ under observation was differentiated twice before being applied to the moving element. Thus the current through the element was

$$i(t) = \frac{d^2 I(t)}{dt^2} \tag{5.4}$$

and by substitution

$$P \frac{d^2 \theta(t)}{dt^2} = G \frac{d^2 I(t)}{dt^2} \tag{5.5}$$

the solution of this equation being simply

$$\theta(t) \propto I(t). \tag{5.6}$$

When the implications of this approach are examined they lead to a design which is quite the opposite of the Blondel technique, namely a relatively large coil mounted on a weak suspension and having a low natural frequency of vibration. Just as in the Blondel system, one cannot be certain of eliminating completely the unwanted terms in the equation and so there is bound to be some residual error from this source in the practical instrument. There is inevitably some damping, however small, and no suspension wire will exert absolutely zero force.

Abraham took this line of reasoning still further, and pointed out that if one applied to the element the current

$$i(t) = p \frac{d^2 I(t)}{dt^2} + k \frac{d I(t)}{dt} + \tau I(t) \tag{5.7}$$

one would obtain the equation

$$P \frac{d^2 \theta(t)}{dt^2} + K \frac{d\theta(t)}{dt} + T\theta(t) = G\left(p \frac{d^2 I(t)}{dt^2} + k \frac{d I(t)}{dt} + \tau I(t) \right) \tag{5.8}$$

and if one can arrange that

$$\frac{p}{P} = \frac{k}{K} = \frac{\tau}{T} \tag{5.9}$$

one obtains *exactly* the desired result that $\theta(t) \propto I(t)$.

Abraham's first rheograph (or 'rhéographe' in the original French) was constructed with the aid of the firm of Carpentier and it endeavoured to use this full compensation technique [41–43]. It consisted of two parts, apart from the usual rotating mirror for time scanning, namely the galvanometer movement itself and what he termed as his 'compensation table', which was a board with various components mounted on it to provide the preconditioning of the current. The circuit of this is shown in figure 5.25. In order to understand its operation, one must first consider the general case of two coils having self-inductances L_1 and L_2 and resistances R_1 and R_2, mutually coupled and having a mutual inductance of M. It can be shown by some elementary circuit analysis that if the inductance of the secondary coil is

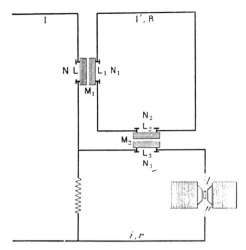

Figure 5.25 Abraham's compensating circuit, as used in the first rheograph [41].

small, the current in the secondary will be approximately

$$i_2(t) = \frac{M}{R_2} \frac{di_1(t)}{dt}. \qquad (5.10)$$

Referring again to figure 5.25; the current to be observed is I, flowing in the left-hand loop of the circuit. L and L_1 constitute a pair of coils obeying the above criterion so that the current I' in the intermediate loop is the first differential of I. Similarly, by virtue of L_2 and L_3, the final current i induced in the galvanometer loop (l,n) will be the second differential of I. If this loop were isolated, this would be the only current flowing through the galvanometer. However the third loop embraces a resistor which is also in the first loop so that a component of current proportional simply to I is added to the d^2I/dt^2 component. The last remaining term, that in dI/dt, was added by mounting the two sets of mutually coupled coils in such a way that there was some linkage of flux between L and L_3. Alternatively, it was possible to introduce this term by including a resistor in common with the intermediate and galvanometer circuits—i.e. in the same way as the I-term was provided. One of Abraham's published papers shows just such a connection [44]. The proportions of the various terms could be varied by using variable slide wire resistors for the coupling and/or by varying the distance between L and L_3. The system was set up correctly by applying a current step to the input, by making the various adjustments referred to above and by varying the current in the galvanometer field coil which altered the galvanometer constant G and had other secondary effects upon the damping (J T Irwin, in 1925, proposed yet another alternative method of compensation using a capacitive circuit [45]).

After careful analysis of the system, the optimum-compromise conditions were found to be as follows.

(i) The resistances of the intermediate circuit and the galvanometer circuit were made equal.

(ii) The self-inductances of the coils were all made equal to that of the galvanometer.

(iii) The magnetic flux of the galvanometer was made as large as possible—i.e. high sensitivity.

(iv) The inertia of the moving coil was reduced by making it tall and thin, and in fact, the moment of inertia of the moving system was virtually that of the mirror which was relatively large to allow for efficient projection of the light beam on to a screen.

The moving system consisted of about fifty turns of fine wire stuck onto the back of the mirror which was concave in shape and had an area of about half a square centimetre. It was carefully balanced so as to avoid movement in unwanted directions. To discourage such movements the suspension was made fairly tight and there were two small cushions mounted behind the torsion wire to prevent backward and forward motion of the mirror and coil.

Professor Blondel was very complimentary towards the rheograph—

The method is, without doubt, the most ingenious of the direct methods and neither inertia nor damping limit its application. [46,47]

However, he did point out the main limitation which was the presence of inductance or 'electrical inertia' which reduced the frequency at which it could operate. He also thought, probably with some justification, that it would be difficult to adjust so as to obtain precisely the correct parameters for complete compensation. He admitted that it seemed to be possible to use it for the observation of frequencies up to 10 000 Hz 'a result out of the question with an oscillograph'. He ventured to suggest that it might be advantageous to combine the Abraham compensation principle with his own bifilar loop since the difficulties in suspending the galvanometer coil described by Abraham in his paper would be overcome by use of the loop. The inductance of the simple loop was also very small. This would allow reduction of the inductances used on the compensation table and hence would extend the upper frequency limit. There was a brief exchange of correspondence between Abraham and Blondel concerning the importance of the initial conditions when considering the solution of the differential equations given above [48]. (Provided that all electrical quantities were initially zero, there seemed to be no problem.)

Judging by the brief account of the rheograph written in *L'Éclairage Électrique* in 1897, the instrument was received with considerable enthusiasm [49]. Abraham gave a demonstration involving the display of four different waveforms in quick succession—a feat which was quite impossible with the lengthy Joubert contact method which was widely used at that time, and with which the audience was familiar.

> It was to the noise of the applause of the audience that Monsieur Abraham projected the curves produced by his rheograph … members of the Society gave to their author a well-merited ovation for his notable work.

A simplified version of the rheograph was announced some ten years later in 1907, again constructed with the collaboration of the firm of Carpentier and exhibited at the Société de Physique exhibition in Paris [50–53]. This reverted to the first simplification of the differential equations described by equation (5.5). As previously mentioned, this involved a relatively large coil on a weak suspension. The diagram of figure 5.26 shows the construction of this new instrument. The moving element was an aluminium ring A, 38 mm tall and 6 mm wide constituting a single-turn coil. A quite large mirror B of area 48 mm^2 was mounted on the ring and the whole was sus-

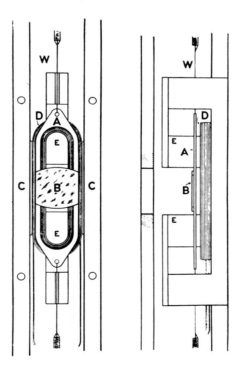

Figure 5.26   Construction of the later, simplified rheograph [51].

pended by the silver wire W in the field of a weak permanent magnet. The ring acted as the secondary winding of a transformer, the primary D consisting of a few turns of wire mounted firmly on the body of the instrument. A laminated iron core E carried the flux through the secondary winding. Since the suspension was weak, the T-term in the equation was negligibly small, and since the field was weak the electromagnetic damping produced by the movement of the coil was also small. The parameters were such that the secondary current flowing in the aluminium coil was proportional to the first differential of the primary current.

Figure 5.27 The complete Abraham rheograph. The moving-coil assemblies (G_1, G_2) are on the left; the prism scanning system is on the right [51].

Figure 5.27 shows a photograph of the complete instrument. The larger assembly on the right is the rotating-mirror system, of which more in a moment. The galvanometer is the smaller item on the left. Two moving-coil assemblies G_1 and G_2 were mounted in the permanent magnet and for this reason it was referred to as the 'double rheograph'. When it was desired to observe a voltage waveform, the voltage was applied to one of the galvanometers through a capacitor. Since the current in a capacitor is given by

$$i_c(t) = C \frac{\mathrm{d}V(t)}{\mathrm{d}t} \tag{5.11}$$

the current in the aluminium coil would be

$$i(t) \propto \frac{\mathrm{d}^2 V(t)}{\mathrm{d}t^2} \qquad (5.12)$$

and equation (5.3) becomes

$$P \frac{\mathrm{d}^2 \theta(t)}{\mathrm{d}t^2} \propto G \frac{\mathrm{d}^2 V(t)}{\mathrm{d}t} \qquad (5.13)$$

to give $\theta \propto V$.

For observation of current waveforms, the current was passed through the primary of a pair of mutually coupled coils arranged as described above so that the secondary current was the first differential of that in the primary. The ring current was then proportional to $\mathrm{d}^2 I(t)/\mathrm{d}t^2$ and from equation (5.3)

$$P \frac{\mathrm{d}^2 \theta(t)}{\mathrm{d}t^2} \propto G \frac{\mathrm{d}^2 I(t)}{\mathrm{d}t^2} \qquad (5.14)$$

so that the deflection this time was proportional to $I(t)$. Both galvanometers could, of course, be used simultaneously to give superimposed curves of voltage and current. It was also possible to display magnetic flux variations by the use of a small search coil.

The mirror unit of the 'synchroscope' unit was rather ingenious. It was, in principle, no more than the traditional rotating mirror, but instead of a plane mirror it used the internal reflections from a rotating triangular glass prism—see figure 5.28. A beam of light from the arc lamp A was reflected from a fixed mirror B onto the moving mirror (or mirrors) of the rheograph and thence into the rotating prism G. If the prism is assumed to rotate in

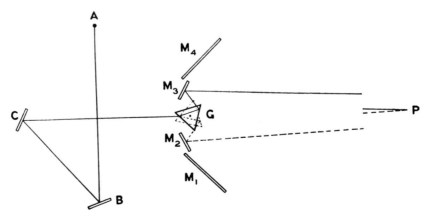

Figure 5.28 Prism 'synchroscope' employed by Abraham to increase the apparent speed of rotation of the scanning mirror [51].

a clockwise direction the incident ray is first reflected from mirror M_1 and sweeps across the screen situated at some distant plane P. It then arrives at the mirror M_2 and is reflected to produce a second sweep, and then in turn from M_3 and M_4. If a simple rotating mirror had been used at G four sweeps would have been produced for one rotation. With the prism, two further internally reflecting surfaces come into play and this produces twelve separate screen scans per revolution, permitting a much lower shaft rotation rate for a given number of sweeps per second. It is equivalent to a drum with twelve mirrors mounted on its sides, but is a much more compact unit. In the photograph of figure 5.27 the toothed wheels and coils of the synchronous motor driving the shaft can be seen at the front of the mirror unit.

This version of the rheograph was, like its predecessor, limited by the inductance of the circuits and also by the incorrectness of the assumption that the current in the ring was exactly the differential of that in the fixed coil. The ratio of inductance to resistance was an important factor in this; the smaller this ratio, the better the performance. With the particular parameters of this instrument it was able to perform accurately up to a frequency of about 2000 Hz. Above this a time lag was produced in the trace but the shape of the waveform was not altered to any appreciable extent. The frequency limit could have been extended by introducing further resistance into the circuit, but the sensitivity would have suffered.

Although the rheographs were undoubtedly very ingenious devices and were well received by the audiences to whom they were first demonstrated, the present author has been unable to determine to precisely what extent they came into regular laboratory use. The Duddell instruments were well established on a commercial basis, as were those of Blondel, and there seems to be no evidence to suggest that the rheographs took any appreciable share of their markets. L T Robinson, writing for the American Institution of Electrical Engineers in 1905 gives an American point of view which may or may not be typical [54]. He commended the rheograph as being 'deserving of special mention' but he does not go into details in his article.

> ... since it was not the intention to touch on anything which was not directly related to practical work. Given conditions to be met which are different from those which prevail at the present time some of these devices might be of the greatest practical use as they have been in special investigations.

In the course of a discussion at the Institution of Electrical Engineers in London in 1907 on the subject of the Irwin hot-wire oscillograph (see Chapter 6) Duddell remarked that the compensating circuits which Irwin had used bore some resemblance to those introduced by Abraham [55]. From Irwin's reply we can infer that the rheographs had not come into general use—'one can only suppose that the method turned out too complicated for ordinary use'.

The rheographs also had the disadvantage that due to their reliance on mutual coupling they were unable to respond to DC inputs and this may have detracted from their attractiveness as general purpose laboratory instruments.

Their designer, Professor Abraham, was a man of considerable distinction who made many contributions, besides the rheograph, to electrical science. He was involved in fundamental electromagnetic research and in the selection of units. Just before the outbreak of the Great War of 1914–18, he went across to the United States and brought back with him to Europe an example of the new triode valve. In collaboration with General Ferrié, the well known radio pioneer, he was responsible for the design of the first thermionic valve to be manufactured in France. It is tragic to have to add, in conclusion, that he died in 1943, aged 75, in a concentration camp [56].

References

[1]   Blondel A 1893 *C. R. Acad. Sci., Paris* **116** 502–5
[2]   Blondel A 1893 *Electrician* **30** 571–2
[3]   Blondel A 1900 *Congrès Int. de Physique* (*Paris*) *1900* vol 3 (Paris: Gauthier-Villars) Report No 8, pp264–95
[4]   Report 1900 *Nature* **63** 142–5
[5]   Blondel A 1901 *Rev. Gén des Sciences* **12** 612–26
[6]   Vigneron E 1907 *Measures Électriques* (Paris: Gauthier-Villars/Masson) vol 2 pp54ff
[7]   Armagnat H 1897 *L'Éclairage Électrique* **12** 346–53
[8]   Armagnat H 1898 *Instruments et Méthodes de Mesures Électriques* (Paris: Carré et Naud) pp567ff
[9]   Colley R 1885 *Wiedemann's Ann.* (New Series) **26** 432–56
[10]   Blondel A *see* [5]
[11]   Blondel A 1902 *L'Éclairage Électrique* **31** 41–50, 161–8
[12]   Blondel A 1900 *GB Patent Specification* 23417
[13]   Hale H 1903 *Electrical World and Engineer* **41** 743–4
[14]   Robinson L T 1905 *Trans. Am. IEE* **24** 185–214
[15]   Gerard E M 1901 *Mesures Électriques* 2nd edn (Paris: Gauthier-Villars) pp466–77
[16]   Report *see* [4]
[17]   Hay A 1907 *Alternating Currents* 2nd edn (London: Harper) pp74ff
[18]   Solier A 1904 *L'Éclairage Électrique* **40** 167–72
[19]   Durand C L 1902 *Electrical Review, NY* **41** 338–40
[20]   Blondel A *see* [11] p162
[21]   Blondel A *see* [3] p282
[22]   Blondel A *see* [5] p624
[23]   Blondel A *see* [3] p294
[24]   Hotchkiss H J and Millis F E 1895/6 *Phys. Rev.* **3** 49–62

[25] Millis F E 1895–6 *Phys. Rev.* **3** 351–8
[26] Millis F E 1896–7 *Phys. Rev.* **4** 128–42
[27] Hotchkiss H J 1899 *Phys. Rev.* **8** 152–60
[28] Report 1899 *The Electrical Engineer* **23** 525
[29] McKittrick F J A 1896 *Trans. Am. IEE* **13** 245–69
[30] Oplinger K A 1933 *Instruments* **6** 193
[31] Laws F A 1938 *Electrical Measurements* 2nd edn (New York: McGraw-Hill) p653
[32] Hanna C R 1925 *Proc. Inst. Rad. Eng. Am.* **13** 437–60
[33] Goldschmidt R 1904–5 *Electrician* **54** 1038–9
[34] Goldschmidt R 1905 *Electrical World and Engineer* **45** 901–2
[35] Editorial 1904–5 *Electrician* **54** 1034
[36] Laws F A *see* [31] p32
[37] Irwin J T 1925 *Oscillographs* (London: Sir Isaac Pitman) pp90ff
[38] Einthoven W 1905 *Ann. Phys.* **16** 20–31
[39] Butterworth S, Wood A B and Lakey A H 1926–7 *J. Sci. Instrum.* **4** 8–18
[40] Martin E J and Caris D F 1932 *Rev. Sci. Instrum.* **3** 598–615
[41] Abraham H 1897 *L'Éclairage Électrique* **11** 145–50
[42] Abraham H 1897 *C. R. Acad. Sci., Paris* **124** 758–61
[43] Armagnat H *see* [7]
[44] Abraham H 1897 *Bull. de la Soc. Int. des Électriciens* **14** 397–434
[45] Irwin J T *see* [37] p53
[46] Blondel A *see* [3]
[47] Abstract 1901 *Science Abstracts* **4** No 1127 p512 (A good summary of [46])
[48] Report of correspondence 1897 *L'Éclairage Électrique* **11** 462
[49] Report 1897 *L'Éclairage Électrique* **12** 180–2
[50] Abraham H 1909 *J. Physique* **8** 265–74
[51] Report 1909 *Electrician* **63** 500–2
[52] Report 1909 *Electrical World* **53** 61
[53] De Coursey H 1907 *Western Electrician* **40** 553 (N. B. The explanation of the operation of the synchroscope is not correct in this paper.)
[54] Robinson L T *see* [14]
[55] Irwin J T 1907 *J. IEE* **39** 643–7 (Discussion)
[56] *see* entry Abraham H in 1960 *Grand Larousse Encyclopédique* vol 1 (Paris: Libraire Larousse)

6

Miscellaneous Methods

Summary

The major lines of development in the science of waveform display have been covered in previous chapters—namely, the contact method in all its forms, and the mechanical oscillographs with vibrating elements. This chapter will describe the various alternative methods which were proposed from time to time. These made use of a variety of assorted physical effects to achieve their ends.

6.1 Chemical methods

In previous chapters of this book we have dealt with those methods of observing waveforms which, by and large, were successful in practice and were widely used in their day, before better methods appeared on the scene to render them obsolete. As might have been expected, knowing the fertile and inventive minds of the nineteenth century scientists and engineers, various other lines of approach to the problem were also suggested. A few of these remained simply as suggestions; for example that of E M Gerard [1] who proposed using a circular band of steel driven by a synchronous motor. This band was to be given an initial magnetisation by passing it near a coil carrying DC current, and then it was to be allowed to rotate between the poles of an AC coil. The waveform would thus be 'written' around the band as an increase or decrease of magnetisation, and this could be measured later by turning the band very slowly near a pivoted magnetic needle, and reading its deflections.

Other lines of approach resulted in practical instruments which were used for a time, but which did not seem to live up to their inventors' expectations for one reason or another. Some of the most promising proposals were those which were referred to collectively as the 'chemical' methods. These relied on the fact that certain solutions undergo chemical reactions and colour changes when they are subjected to the influence of electrical voltage and current. For example, if a piece of paper is soaked in a solution of potassium iodide then application of a potential difference to two electrodes touching the paper will cause a brown iodine stain to appear at the anode. This was applied practically in 1864 in a simple communication system known as Bain's telegraph [2–4]. This used the Morse code, the dots and dashes being represented by pulses of DC voltage on the line. At the receiver, a strip of paper soaked in the potassium iodide solution was passed over a metal roller. A stylus was pressed on the strip and the incoming signal applied between the stylus and the roller. The signal was thus recorded as a brown line of dots and dashes on the strip. The principle of recording in this way was also revived some years later by Delaney [5].

The sensitivity of the system could be improved by adding starch to the potassium iodide. A mixture of one part potassium iodide to 20 parts starch paste to 40 parts water was said to be suitable. An even more effective recording medium was made up of equal parts of a saturated solution of prussiate of potash (potassium ferricyanide) and a saturated solution of ammonium nitrate, diluted with two volumes of water. Ammonium nitrate is a deliquescent substance which was added to keep the paper permanently damp. If iron or steel styli were used, the solution reacted at the anode to form Prussian blue, a particularly bright and clear trace being formed in this way [6]. With all these various mixtures, the extent of the reaction, and hence the darkness of the trace, depends on the voltage applied, or alternatively expressed, on the current flowing through the paper.

In 1887, P Grützner, who was a physiologist by profession, attempted to record current waveforms in this way in the course of some experiments to determine the effects of electric impulses on nerves and muscles [7]. He returned to this topic again and on that occasion, in 1899, investigated the behaviour of some electrical circuits, one of his experiments being illustrated in figure 6.1 [8,9]. D is a Daniell cell which is connected across a coil of wire Sp through a switch C. The ends of the coil are in contact with two platinum electrodes E and E_1 pressing onto a moving strip of iodide–starch paper. The sketch on the right-hand side shows the waveform of the voltage across the coil as C is closed and opened. When it is closed, a current is established in the direction shown by the tail-less arrows and after an initial overshoot it settles to a steady value. When the switch is opened there is a kick of reverse voltage which drives current through the paper as shown by the arrows with tails. The paper is moving in a direction into the diagram— i.e. in a direction at $90°$ to the line joining E and E_1. Since the darkening of the paper is proportional to the voltage, the traces produced will be of the form shown beneath the waveform. Figure 6.2 shows a photograph of various traces produced by Grützner in the course of his experiments, that labelled II Sp corresponding to the situation just described.

Another type of electrical waveform which he observed was that

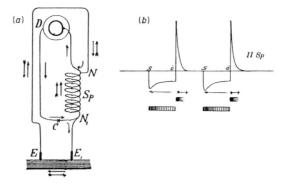

Figure 6.1   (*a*), Grützner's circuit for the investigation of transient currents in an inductor. The sketch (*b*) shows the current waveform and (below it) the form of the trace expected [8].

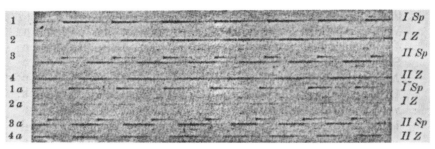

Figure 6.2   Traces produced using the apparatus of figure 6.1. Those labelled II Sp correspond with the one sketched on the previous diagram [8].

generated by a primitive alternator consisting of a magnet rotating near a stationary coil. The type of waveform produced is shown in figure 6.3, the type of trace expected being shown diagramatically just beneath it. The actual traces are seen in figure 6.4, the various periods being the result of rotating the magnet at different speeds. In order to produce a mark, the applied voltage has to be above a certain threshold value. The bottom trace in particular clearly exhibits the gaps where the voltage falls below this level in each cycle.

We have here a situation in which the voltage at any time is represented by the darkness of the trace. Clearly the amount of quantitative information available from such traces is very limited. It is interesting to note that displays of this general sort are still in use today in the sonograph instruments for the spectral analysis of speech and music signals [10].

Although an improved type of dry recording paper marked by a spark is now used (teledeltos paper), signal amplitude is still represented by blackness of trace, and there is still the problem of threshold and of limited dynamic range between full black and full white.

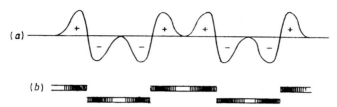

Figure 6.3 (*a*), Voltage waveform produced by rotating a permanent magnet near a stationary coil, and (*b*) the type of trace which Grützner's method would be expected to produce [8].

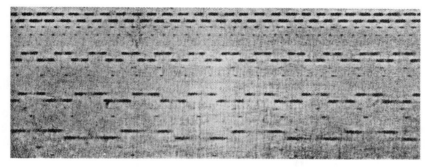

Figure 6.4 Actual traces produced by rotating the magnet at various speeds. These correspond very well with the trace sketched in figure 6.3(*b*) [8].

An improvement was made to this method by P Janet in 1894. This is illustrated in figure 6.5 [11–16]. Assume that a stylus is pressing onto a metal drum around which a piece of iodide paper is wrapped. A sinusoidal voltage S is applied between the stylus and the drum. In order to make a mark on the paper, the voltage must exceed some threshold level which is marked X on the diagram—this would typically be about 1 volt. The stylus will thus leave a record of dashes on the paper as the drum rotates, the lengths of the dashes corresponding to AB, A'B' etc. Now let a positive DC bias voltage be applied to the sinusoid so that it moves upwards, becoming the curve S'. The waveform is now above the threshold for a longer time than before, and the lengths of the dashes will be C_1D_1, $C_1'D_1'$. If we start off with the first trace AB, A'B' etc then we can draw beneath it the points C and D because we know the voltage C_1C, D_1D which was the bias voltage applied; we also know the lengths AC_1 and BD_1 by comparison of the two traces. Having thus established the positions of points C, D etc., the process can be repeated with other positive and negative values of bias and at each stage other points such as C and D can be added so that gradually a picture of the waveform can be built up.

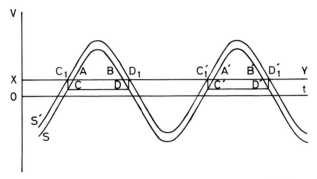

Figure 6.5 Illustrating the principle of Janet's chemical inscription method. This graph shows voltage as a function of time [14].

This was clearly a very tedious procedure, and Janet went on to build a piece of apparatus which would produce the waveform directly without the need for computation of each point. The single stylus was replaced by a number of different styli (which were actually steel sewing needles), each of which was biased by a different amount by means of batteries as shown in figure 6.6. This diagram shows eight needles only, but in Janet's apparatus

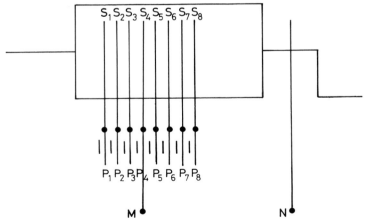

Figure 6.6 Janet's apparatus for the automatic inscription of waveforms by electrochemical means. (P_1, P_2 etc represent the bias batteries) [14].

there were fifteen. The voltage waveform under observation was connected between points M and N. The short lines between the styli in the diagram are meant to represent the bias cells P_1, P_2, each of 4 V EMF. The result is that when the drum is rotated, each stylus draws a line of different length as shown in figure 6.7, and the actual waveshape can be seen as an outline or silhouette of the lines.

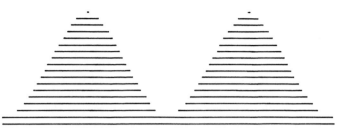

Figure 6.7 Type of trace produced by the apparatus of figure 6.6 [14].

This is quite a neat method of determining the waveform, but suffers from the disadvantage that one has to provide 14 bias batteries. Blondel [17,18] suggested doing away with all these batteries by the use of the circuit of figure 6.8. Here the voltage between the two lines A_1B_1 and A_2B_2 is to be observed. One value of the bias voltage is used and the waveform is added to this in various proportions, once again producing lines of unequal lengths on the paper. Blondel worked out values for the various resistors

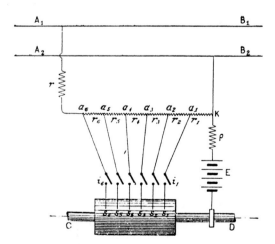

Figure 6.8 Blondel's suggestion for replacing the batteries shown in figure 6.7, with a resistive potential divider adding the waveform under observation to a DC voltage in various proportions [18].

in the circuit. However, the analysis he gives is very dubious and a little thought will show that the component values required will actually depend on the shape of the waveform being observed, so distortion is inevitable. One wonders why he did not think of the easier solution of using a chain of resistors to provide the different bias voltages. Provided that the resistors in the chain were of low value compared with the resistance of the paper between styli, this would have worked quite well.

This method of 'chemical inscription' as it was often called was also used in the determination of frequency and phase [19–21]. A single stylus to which the AC waveform was applied was used to produce a dotted trace as previously described, and the speed of travel of the paper beneath the stylus was determined by using a second stylus activated by a voltage which was switched on and off by the swinging of a seconds pendulum. By comparison of the two traces, the frequency could be deduced. If the applied voltage is sinusoidal in shape, the centre of the dashes marks the instant of maximum voltage. By comparing the traces produced by two alternating voltages applied to two styli, the relative phases between them can be measured. W König attempted to use a somewhat similar technique, but instead of the chemical marking method, he made use of what are sometimes called Lichtenberg figures [22–24]. A stylus laid down a pattern of charge on a moving plate of bitumen, resin, varnish or some such insulating material, and this was revealed by sprinkling on a mixture of fine red-lead powder and sulphur, the red lead clinging to areas of negative charge and the sulphur to positive areas. The traces could also be revealed by laying down the patterns of charge onto a photographic plate and developing it in the normal way, a phenomenon investigated by J Brown in 1888 and again in 1896 [25,26]. These do not seem to have been very successful however, and König himself [27] seemed to be of the opinion that both his own method and that of Grützner were of rather limited application, as indeed they were.

The idea of using Lichtenberg figures on a photographic plate was revived again in the 1920s by J F Peters of the Westinghouse Company in the form of the 'Klydonograph', (from Klydono; a surge or billow) an instrument for recording voltage surges on transmission lines [28,29]. This device consisted essentially of an electrode with a spherical end pressing on a photographic plate backed with a metal sheet. It could be connected to the transmission line in a number of different ways. When a voltage surge due to a lightning flash or a fault condition appeared on the line, the resulting flashover caused a corona-like picture to appear on the plate when developed. By careful examination of this trace, it was possible to determine the polarity of the surge, whether it was steepfronted or gradual, etc. In one version, the photographic plate was made circular and was rotated by a clockwork motor so that the time of occurrence of the fault could be established. The klydonograph was manufactured by the Westinghouse Company for a number of years; for example, it appears in the 'Instruments and Relays' section of their 1930 catalogue. The price of the simple version (without clockwork movement) was $50. The price of the motor-driven version is not stated, and only a passing reference suggests that such a version was available at all. To quote the catalogue: 'By reason of the low cost and reliability it is the logical first step in the investigation of any electrical circuit where abnormal voltage conditions are suspected'.

An interesting variant on the electrochemical method is described in a US Patent granted to E J Murphy in 1902—see figure 6.9 [30,31]. In this

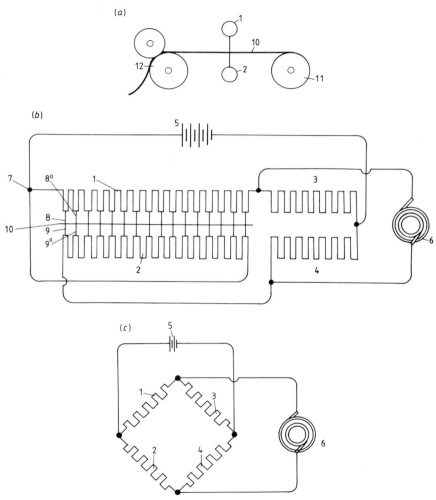

Figure 6.9 Murphy's method of chemical inscription: (*a*), side view; (*b*), front view (schematic); (*c*), electrical circuit [30].

device a thin band of iodide–starch paper is caused to pass between two sets or 'combs' of styli, one on top of the band and the other directly opposite on the underside. The paper band is marked 10 in the side and front views of figures 6.9(*a*) and (*b*). The styli are connected at regular intervals along two resistors, marked 1 and 2 and these, together with 3 and 4, form a bridge circuit connected to the battery 5 of figure 6.9(*c*). Assume for simplicity that all the resistors are equal in value, that point 7, the junction of resistors 1 and 2 is at zero potential, and that the battery voltage is V_b. The distributions of voltage across the resistors will then be as illustrated in figure 6.10(*a*), taking account of the fact that the current flows from left to right in 1 and from right to left in 2. Every opposing pair of styli will thus have a potential difference between them save for the two centre ones which will be at the same potential. Under these conditions, if the paper is drawn through past the points each pair will make a dark line except the centre pair. Since the darkness only appears at one electrode, half of the lines will appear on top of the paper, the rest on the bottom surface. However, the paper is very thin so that all traces can be seen clearly from either side. Now let another potential waveform be applied across the other corners of the bridge circuit. Figure 6.9 shows an alternator 6 connected in

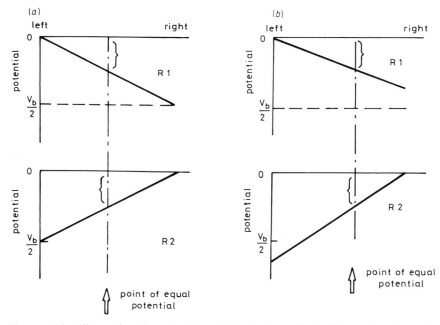

Figure 6.10   Illustrating the principle of Murphy's method. The graphs show the distribution of potential across the rows of styli: (*a*), with no applied signal voltage; and (*b*), with applied signal voltage.

this way. If at some instant of time the top terminal of the alternator is positive this will superimpose a current which opposes that in resistor 1 but adds to that in resistor 2 so that the potential distribution is altered to that shown in figure 6.10(*b*). The position where the opposing styli have no potential difference between them is now shifted from the centre and the 'no trace' line on the record will also have moved. If an alternating voltage is produced by the generator 6 then the 'no trace' position will oscillate across the paper strip and a trace such as that shown in figure 6.11 will result, the waveshape being revealed by the gaps in the lines.

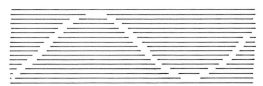

Figure 6.11   Type of trace produced by Murphy's apparatus.

There is rather an odd footnote to the story of the chemical inscription methods. In 1897, W Duddell wrote a short article in the *Electrician* in which he surveyed the whole range of various proposed oscillographic techniques [32]. When dealing with Janet's method, he described what is clearly the arrangement of figure 6.6, but he explains it in these terms;

> at any instant there will be a style for which the algebraic sum of the PD's between it and the cylinder is approximately zero, and this style will not make a mark; by rotating the cylinder we thus get the curve by the gaps in the series of parallel straight lines which the styles draw.

He seems to have misunderstood the principle of operation of Janet's method, but had anticipated that of Murphy by four years or so.

To sum up the chemical methods of waveform inscription; one gets the impression from the literature of the time that perhaps Janet's method was occasionally used for a few years although it never achieved anything like the popularity of the point-to-point and oscillographic methods, and the present author has never seen any curves recorded in this way reproduced in the journals. It would certainly have been difficult to determine the exact positions of the ends of the marks on the moist paper since the stains would have diffused slightly through the solution, and the method would always have been subject to this inherent lack of precision. No doubt such a system would also have been rather messy to use.

6.2   Optical systems using polarised light

Another wave-tracing method, which, for a brief period, seemed to hold out considerable promise was the optical system developed by A C Crehore in 1895 [33]. This made use of Faraday's discovery that if a beam of polarised light is passed through certain substances or liquids which are subjected to a magnetic field, the lines of force of which are parallel to the light beam, then the direction of the polarisation of the light is rotated. The effect was further investigated by Vedet who found that the amount of rotation produced was given by

$$\theta = \alpha_V HL$$

where H is the magnetic field strength in the direction of the beam, L is the distance through which the light travels in the substance and α_V is a constant of proportionality known as Vedet's constant. The value of α_V is a function of the particular substance used, being especially large for carbon bisulphide, ethyl iodide and benzene. The above formula holds for light of any given wavelength, but the rotation also varies with wavelength, being approximately proportional to $1/\lambda^2$ [34].

Both D'Arsonval and Bequerel attempted to use this effect for the measurement of current—the so-called 'optical ammeter' [35,36]. Crehore's apparatus is shown in figure 6.12. Sunlight from a heliostat mirror was first passed through a Nicol prism b which polarised it in one direction. It was then transmitted through a tube d containing carbon bisulphide, finally passing through a second Nicol prism e. The function of the rest of the apparatus will be described in a moment, but first consider what would happen if the light was filtered so as to be virtually monochromatic before passing through the apparatus. The second prism could be rotated to the 'cross-polarised' position so that no light emerged at the right-hand side of the apparatus. If a current were now to be passed through the coil wound around the tube the resulting magnetic field would alter the polarisation plane and some light would emerge from the second Nicol prism. Rotation of the second prism to produce extinction again would permit a measurement of the magnitude of the current; this was the principle of the optical ammeter.

What would now happen if *white* light were put into the apparatus and a current passed through the coil? Since the rotation depends on the wavelength, different components of the white light would be rotated by different amounts. If the second Nicol prism were now rotated, a series of brilliant colours would be seen, the colour at any particular angular position being the complementary colour to that which happened to be cross polarised in that position. If the second prism is set in one angular position and the current through the coil is varied, then at any one particular value of current one component of the light would be extinguished, and colours would again be seen. If the light output is split up into its components by means of an ordinary glass prism or a diffraction grating then a spectrum will appear, but there will be a gap in that spectrum where the absent colour would have fallen. Moreover, as the current in the coil is varied, the gap or dark line in the spectrum will move to and fro along the spectrum. If the spectrum is allowed to fall on a moving photographic plate, the current variations will be recorded as a variation in position of the dark line.

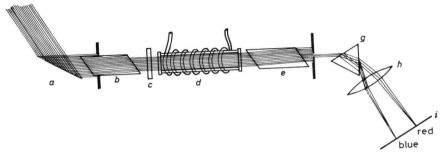

Figure 6.12 Crehore's optical method of curve tracing [33].

One further feature is necessary if the scheme is to work properly. When the current in the coil is zero, all wavelengths are rotated by the same amount and no dark band appears so that the centre of an alternating waveform would be missing. A thin quartz plate c inserted in the light path restores differential rotations between the wavelengths. The output Nicol prism is rotated so that the dark band appears in the centre of the spectrum for zero current and an alternating current in the coil will then produce a movement of this band up and down the spectrum.

Because of the fact that the rotation of polarisation is proportional to $1/\lambda^2$, it would seem that the relationship between the position of the dark band and the current flowing in the coil is not a simple one. The situation was analysed in some detail by Dr Crehore and he showed that reasonable linearity could, in fact, be achieved if a diffraction grating were used to produce the spectrum instead of the prism.

This is really quite an ingenious idea. The movement of the dark line over the spectrum looks like the realisation of the dream of all meter designers, namely a truly weightless, inertialess pointer. However, as always, there is another factor to take into account. In order to produce reasonable movement of the dark band, a fairly high magnetic field strength is needed. This implies a coil of many turns wound over the tube, and such a coil possesses a large self-inductance. In other words, we have exchanged mechanical inertia for its electrical counterpart. Another disadvantage is that the dark band is not very sharp. One particular wavelength of the light may be extinguished completely, but its near neighbours in the spectrum are also diminished. Thus the image of the band formed on the photographic plate will be fuzzy edged and ill defined.

The paper in which Dr Crehore announced this method of waveform recording was confidently entitled 'A reliable method of recording variable-current curves'. When one reads the paper, one realises that it is almost entirely theoretical and speculative in nature. J Blondin, the Scientific Director of the journal *L Éclairage Électrique*, reviewing Crehore's paper pointed out several possible disadvantages [37]. Photographic plates of high sensitivity would be needed, and he doubted whether those available commercially would be up to the job. He also feared that temperature variations would prove to be a nuisance. He wrote, rather scathingly

> It would be useful if the author, instead of describing a method employed already by a number of others for other purposes, had given us the results of his experiments which, according to him, promise much for the future.

Wm Duddell also seemed to be very dubious about the practicability of the method when writing the general review paper previously mentioned [38].

Blondin's comment about other experimenters is possibly a reference to the work of Monsieur J Pionchon of the University of Grenoble which was published at about the same time [39,40]. His apparatus also consisted of a tube filled with either carbon bisulphide or else a mixture of mercury iodide and potassium iodide, surrounded by a solenoid carrying the current under observation. When an alternating current was passing through the coil, the light output of the second Nicol prism would fluctuate, but when viewed directly by eye there would be no visible effect due to persistence of vision. However, if viewed stroboscopically with a flashing light the variations could be slowed down for further study. The method would only work properly for monochromatic light, and to obtain this Pionchon passed white light from an arc lamp through a solution of potassium bichromate. The stroboscopic flashes were produced by two diaphragms mounted on the prongs of an electrically maintained tuning fork, each diaphragm having a narrow slot cut into it. A flash was produced each time the slots came into alignment. Henri Abraham, commenting upon this, expressed the opinion that there would be two main disadvantages with the method. The first was the difficulty in obtaining a monochromatic light source of sufficient intensity; the second was that the flashes of light produced by the tuning fork arrangement were of rather long duration and so the resolution of the current waveform would be limited.

Abraham, assisted by Monsieur H Buisson, proposed an improvement on the method which alleviated these problems [41,42]. First they illuminated the tube by means of a spark from a Leyden jar capacitor, the high voltage being produced by an induction coil whose primary contact was operated by a switch mounted on the axle of the alternator whose waveform was being examined. In this way a short flash of light could be made to occur at the same point in each cycle of the waveform. This flash was viewed through the tube of mercury iodide/potassium iodide solution and two

Nicol prisms which had been crossed so as to produce extinction of the light under no-current conditions. When the alternating current was applied, the rotation of polarisation meant that light was again seen, the extent of the rotation being dependent on the instantaneous value of the current at the time of the flash. This was counteracted by passing a DC current through another coil wound on the tube in such a sense as to restore the rotation to zero. The value of this DC current, read from a meter, then gave the value of the AC current at that instant. The timing of the spark throughout the AC cycle was then varied in the manner of a Joubert contact and a point-by-point picture of the waveform was built up. This would undoubtedly have worked after a fashion, but the inductance of the coil was still a complicating factor, the tube was undoubtedly messy to set up and the whole system was really overelaborate for what it actually achieved. One could as well have used the simple Joubert method and been done with it.

Crehore's system was taken up by J A Switzer, who, in 1898 [43], published his results in a paper which was rather provocatively given precisely the same title as that used by Crehore himself—'A reliable method of recording variable current curves'. He continued the theoretical analysis

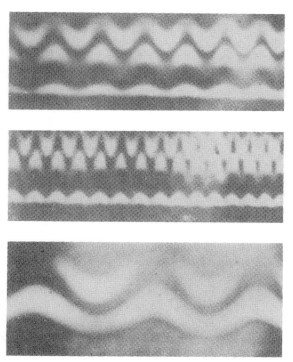

Figure 6.13 Alternating current traces obtained by Switzer using Crehore's method [43].

of the method and showed that not only did the position of the dark band in the spectrum vary with the value of the current, but its width varied also. He carried out some practical experiments and published the photographs shown in figure 6.13. These show the curves obtained for alternating currents of various frequencies. His results are best summed up in his own words—

> These photographs are selected from many as being the best that were obtained. That results very much better than these could be obtained by this method, the writer doubts; and, while the method presents itself as one full of attractiveness, its limitations are necessarily so great that it is not likely to find practical application—at least to the recording of variable electric currents. Dr Crehore has however applied the principle, with signal success, to quite another problem, that of recording the velocity of projectiles [44]; and it may be that still other fields of usefulness are yet awaiting magnetic rotation of polarized light.

The method sank into obscurity and was heard of no more.

6.3   The mercury jet

Another idea which saw the light of day briefly in 1893 was the mercury jet method proposed by E L Nichols [45–47]. In a preliminary experiment a thin stream of mercury was allowed to fall vertically from a jet. It fell between the electrodes of an electrostatic machine and was thus deflected sideways. This deflection of the jet was observed as the machine alternately charged and discharged. It turned out to be a very insensitive system and was replaced instead by the arrangement of figure 6.14. The thin jet of mercury fell between the poles of a permanent magnet. When a current was passed through the jet it experienced a sideways force, and hence a deflection. The jet was viewed through an illuminated horizontal slit and its movement was photographed, appearing as a black line on a clear background when turned into a positive print. Nichols' paper contains a very convincing sinusoidal curve produced in this way. It must be remembered that movement of a jet of liquid is very different from that of a string galvanometer although, at first sight, it would seem to be very similar. In the case of the jet, it experiences a deflecting force, but there is no restoring torque acting simultaneously to bring it back to the centre position. The consequence of this is that it has no natural frequency of its own. Blondel expressed his surprise that Nichols had obtained curves which, apparently, represented the current satisfactorily using such an instrument [48].

Some years later, J T MacGregor-Morris and R Mines [49] carried out a careful analysis of a jet system and showed that it would indeed be possible to produce reasonable traces in this way and that the response to high-frequency signals could be made satisfactory for many purposes. Nichols

seems to have been prompted to try out this method by the difficulties he
had experienced when trying to use the Frölich diaphragm system of
waveform recording (figure 2.44). As he said at the American Association
meeting in 1893,

> Some experience with Frölich's method has convinced me that not only is
> extraordinary skill necessary in order to obtain, by means of a mirror attached
> to the diaphragm of a telephone, curves which should represent even with a
> fair approximation the law of whatever periodic changes we may desire to
> record, but that attainment of the proper adjustment is a matter so entirely
> fortuitous and its maintenance so uncertain as to deprive the method of much
> of its usefulness. ...
> The remedy clearly consists in the elimination of mechanism and in reducing
> the inertia of the moving parts.

Nichols did not develop the method beyond the purely experimental
stage, and little more was heard of it.

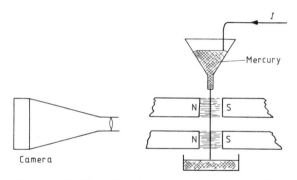

Figure 6.14   The Nichols mercury jet system [45].

6.4   Electrostatic and hot-wire oscillographs

An oscillograph working on electrostatic principles was devised by E Taylor
Jones in 1907 [50]. This can be seen in figure 6.15. A horizontal phosphor-
bronze strip A is kept in a state of tension by means of a helical spring. To
the rear of the strip is a metal plate D whilst at the front, embedded in a
block of ebonite F, is a further plate E. The plate D is in electrical connection with the bronze strip. When a potential difference is applied between
the plates, the strip is attracted towards the plate E. This movement is
observed by means of a small triangular mirror fixed to the centre of the
strip, its upper end resting against the ebonite rod C. A light beam for
observation of the movement is admitted through the window G, a slit being
cut into plate D to allow uninterrupted passage of the light to and from the
mirror.

This instrument was clearly most suited to the observation of high-voltage waveforms. The version of it which is shown in figure 6.15 is that produced by the Cox Cavendish Electrical Company and exhibited by them at the Physical and Optical Society in 1922 [51]. There are several features of interest to be noted. Firstly, since there was no polarising voltage, the direction of movement was independent of the polarity of the applied voltage under observation. Bipolar or alternating voltages appeared in full-wave rectified form when displayed with the aid of the usual rotating-mirror system. Secondly, the deflection was proportional to the square of the applied potential difference and this had to be taken into account when interpreting the waveforms produced. The cavity of the instrument was filled with oil which served for both insulation and damping; a mixture of equal parts castor oil and Singer's machine oil was recommended. The sensitivity could be varied by plugging in different ebonite rods F having attracting plates E of differing sizes set at different distances from the strip. Zero setting of the trace was obtained by raising and lowering the rod C on which the upper point of the mirror was resting. The frequency response was not stated directly in Taylor Jones's paper but he says that 'it could resolve waves having periods of .00488 seconds and .00238 seconds', from which we can imply that it was usable up to frequencies of about 400 Hz.

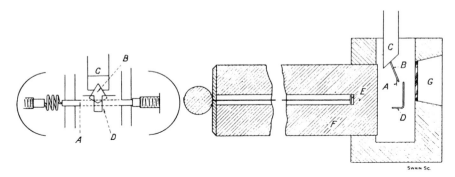

Figure 6.15   The Taylor Jones electrostatic oscillograph [51].

The electrostatic principle was developed further by Professors Ho and Koto of the Tokyo Imperial University in 1913 [52–55]. The arrangement of their oscillograph is shown in figure 6.16(*a*), the principle of operation being illustrated in figure 6.16(*b*). It is assumed that the potential difference variations between points a_0 and b_0 are to be displayed, and that this is rather too large to be applied directly to the instrument itself. A capacitive voltage divider chain is thus provided to reduce the voltage. This could be proportioned so as to suit any particular circumstances, but it will be assumed here that the two capacitors C_1 and C_2 are nominally equal in value. This reduced potential is applied between the two plates F_1 and F_2.

The two bronze strips S_1 and S_2 are mounted vertically under tension in the manner of the Blondel bifilar oscillograph, but here the strips are insulated from each other by a silk thread passing over the pulley. A battery of about 300 V connected as shown maintains a potential difference between the strips. The windows w in the plates allow for observation of a small mirror fixed to the strips.

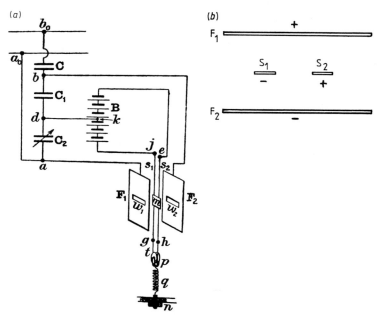

Figure 6.16 Part (*a*) shows the electrostatic oscillograph of Ho and Koto. Part (*b*) shows a plan view of the moving strips s_1, s_2 situated between the plates F_1, F_2 [52].

If it is assumed for simplicity that the potential at the point d is zero, then the potentials at the plates and strips will be as shown in figure 6.16(*b*). (If d is not actually at zero potential, this will simply add a constant to all the other potentials with no overall effect.) The effect of this potential distribution is that one strip will be attracted forward, and the other back so that a torque is produced on the mirror as in the Blondel oscillograph. Reversal of the voltage between a_0 and b_0 will reverse the direction of the torque. Detailed consideration of the electrical fields and forces in the instrument leads to the conclusion that the deflection will be directly proportional to the potential difference between a_0 and b_0 so that the squaring effect of the unpolarised Taylor Jones instrument has been eliminated.

For proper linearity the strips must be situated precisely mid-way between the two field plates, but this proved to be very difficult to achieve in practice.

However it could be compensated for by a slight unbalance in the potential divider chain, which is why C_2 is shown as a variable capacitor.

The actual plate/vibrator assembly is shown in figure 6.17. As an indication of scale, the field plates were 9 mm wide and 5 mm apart. When in use it was immersed in oil for the usual reasons. It was made commercially by the Cambridge Scientific Instrument Co. and as the authors remark 'the vibrator has approximately the same dimensions as the latest Duddell vibrator (cf figure 4.22). These two types of vibrator can thus be placed conveniently side by side in the same oil bath, for which purpose one of the field plates F_2 is made of soft iron to serve as part of the magnetic circuit'. The authors state that voltages up to 9 kV had been measured with this instrument, and the resonant frequency was just over 3000 Hz. It was clearly most suited to the measurement of voltage waveforms but they indicated that it was also usable for current waveforms by observation across a resistor of suitable value.

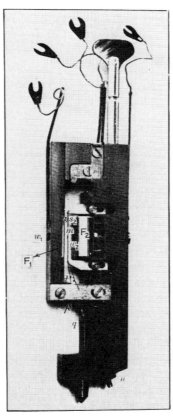

Figure 6.17   Version of the Ho and Koto oscillograph produced by the Cambridge Scientific Instrument Co. [52].

Another device employing the bifilar suspension principle was the hot-wire oscillograph invented by J T Irwin in 1907 [56–59]. This was an adaptation of the well known hot-wire ammeter principle. If a current is passed through a piece of fine wire, the heating causes the wire to expand and the extent of the expansion can be used as a measure of the magnitude of the current. Perhaps the best remembered of these hot-wire ammeters was that of Cardew [60] dating from 1884, although similar meters were produced by other people including M Mayençon [61], Ayrton and Perry, and Johnson and Phillips [62]. Since the heating is proportional to the square of the current these instruments had a very non-linear scale, cramped at its lower end.

The principle of the Irwin oscillograph is illustrated in figure 6.18. Two fine wires, CF and DE are held rigidly at their ends. A polarising battery B_1 sends equal currents b through each wire so that both are initially heated to the same temperature. If an external current a is caused to flow as shown, a proportion of that current will flow through the wires, opposing b in the left-hand wire and aiding it in the other. If the resistance of each wire is r it is easily shown that the power dissipated in the left-hand wire is

$$\left(a \, \frac{R_1}{R_1 + r} - b \right)^2 r$$

that in the right is

$$\left(a \, \frac{R_1}{R_1 + r} + b \right)^2 r.$$

The difference is

$$4br \, \frac{R_1}{R_1 + r} \, a$$

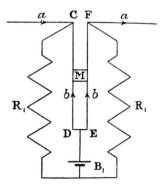

Figure 6.18 Illustrating the principle of operation of the Irwin hot-wire oscillograph [56].

i.e. it is proportional to the external current *a*. In other words, if a difference in expansion between the two wires is observed this serves to remove the square-law nature of the device and render it linear in operation. Irwin initially tried to measure this differential expansion by fixing fine springs to the centres of the two wires so that they were held under tension, and attaching a small mirror across the wires in the usual way, a difference in sags thus resulting in an angular tilt of the mirror. He experienced a great deal of difficulty and frustration with this method and was unable to equalise the spring tensions to an accuracy better than about 10 %. He then abandoned this simple springing for more complicated systems of crossed loops tied together in diagonal fashion as shown in the example of figure 6.19. This was a self-equalising system which greatly eased the difficulties. The graph published by Irwin showing the deflection as a function of the current *a* is impressively linear. The actual vibrating element can be seen in figure 6.20.

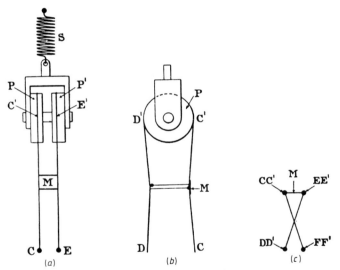

Figure 6.19   The improved cross-connected double-loop system of Irwin's hot-wire oscillograph: (*a*), front view; (*b*), side view; (*c*), plan [56].

The principle of operation of the hot-wire oscillograph sounds very simple, but in reality its action is rather complex. When a current is passed through a wire, the wire starts to heat up and continues to do so until the heat generated by the current is equal to that lost by convection, radiation and conduction. This means that some appreciable time must elapse after the initial application of the current before a steady deflection is attained. The rate of change of current which the instrument is able to follow depends on the ratio of the quantity of heat stored in the wires for a given rise in temperature to the increased heat lost for the same temperature rise. The rate of heat loss, and hence the speed of response, can be increased by immersing the wire in oil but even so, according to Irwin, in this form, it would not be possible to use the system for frequencies greater than about 5 Hz. Irwin compensated for this lag in response by introducing a capacitor–resistor combination into the circuit when voltage waveforms were being observed, and mutually coupled coils when current waveforms were required. This technique is similar in many ways to the Abraham rheograph techniques described in the previous chapter.

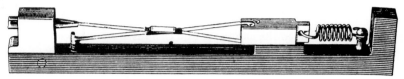

Figure 6.20 The vibrating element of the Irwin hot-wire instrument [56].

In Irwin's oscillograph, the mirror fixed to the wires was fairly large (4 mm × 1.5 mm) since it was required for projection purposes, but even so, the natural frequency of oscillation was about 3200 Hz in air (2360 Hz in oil) so it was quite suitable for use with mains frequency currents and voltages. He discovered that the tension in the crossed tie wires was more important than the actual wire-loop tension as regards pushing up the resonant frequency, and by increasing this cross tension and using a smaller mirror he believed it would be possible to attain a resonant frequency (in air) of up to 20 000 Hz. The sensitivity, using castor oil as the cooling–damping fluid was such that 0.5 A flowing in the wires produced 30 mm deflection on a scale situated 1 m away.

In the discussion which followed the presentation of his paper at the IEE several contributors, Duddell included, expressed their astonishment that he could have adapted the hot-wire principle, which looked so unpromising, to the observation of waveforms and they offered him their warm congratulations on his ingenious instrument. When Duddell had published his own paper on the oscillograph, he included, as an illustration of its capabilities, some waveforms associated with an alternator. Irwin had been able to gain access to the very same machine and in his presentation he showed the waveforms obtained with a Duddell oscillograph alongside his own, thereby demonstrating that since they were virtually indistinguishable, his method was capable of giving excellent results.

6.5 The electrocapillary method

A very elaborate method of waveform determination was suggested by G J Burch in 1896. Burch, it may be of some interest to note, was also responsible for the introduction, in a letter to the *Electrician* in 1906 (p 602), of the well known rule for the magnetic field direction in a coil, represented by the sketches

His waveform display apparatus was based on the type of electrometer introduced by Lippmann in 1874 [63–65]. The principle of operation is shown in figure 6.21. A fine capillary tube containing mercury in its upper part dips down into a bath of dilute sulphuric acid, being so assembled that there is no air bubble in the capillary, and there is a mercury/acid interface as shown. If an electrical potential difference is now applied between the mercury and the acid—this contact being made via a pool of mercury in the acid—surface tension changes occur and the interface moves up and down the tube. The system is very sensitive, and it was claimed by Burch that a movement corresponding to as little as 1/30 000 volt could be observed with the aid of a microscope. The method was quite widely used for some years, particularly by Physical Chemists and Physiologists for measurement of electrode potentials etc [66,67].

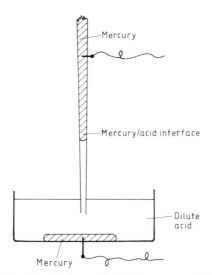

Figure 6.21   Simple form of the Lippmann electrometer.

The idea behind Burch's method for waveform determination was that the waveform should be applied across the electrometer, the resulting movement of the mercury being projected onto a moving plate and recorded photographically. As we shall see later, there are several complicating features which have to be accounted for if the method is to succeed.

In his published paper [68], Burch gave elaborate instructions regarding the construction of the apparatus and, in particular, how the capillary was to be set up. He tried several forms, the one finally recommended being shown in figure 6.22. The capillary is formed in a thick-walled glass tube, one half of which is ground away so that a trough is created. This is closed with a thin flat plate of glass so that a mercury/acid column of D-shaped cross section is formed. In this way the capillary can be illuminated from behind and photographed without the distortion that a cylindrical tube of the usual sort would have produced.

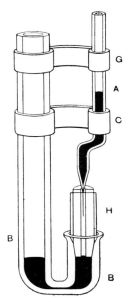

Figure 6.22   Form of Lippmann electrometer developed by Burch for purposes of optical viewing and projection [68] (p382).

In his first model, movement of the photographic plate was provided by the hydraulic motor shown in figure 6.23. A is a horizontal brass tube slotted at the top, and inside this tube there are two pistons B connected by a thin hollow rod in the centre of which is the tap C. If this tap is open, water can pass freely through the rod and it can be moved from side to side within the outer tube. The tap G at the left is connected to a mains water supply. If the small tap C is closed, the pressure of the water causes the whole thing to move from left to right. A photographic plate is affixed to a slider D which is in turn fastened to the rod. To use the device, the two mains water taps F and G are closed and the tap C is opened. The plate holder can then be moved easily to the left, C is then closed. Tap F is opened and then G is opened suddenly, the result being a steady movement of the plate to the right. The speed of movement is controlled by the degree of opening of the tap F, speeds in the range 0.3 to 5 cm s^{-1} being attainable. Incidentally, care had to be taken to see that the outlet tube from F was always under water to prevent air from being drawn back into the apparatus when setting it.

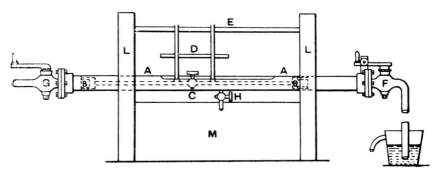

Figure 6.23 'Hydraulic motor' constructed by Burch for imparting uniform horizontal motion to a photographic plate [68] (p435).

This arrangement was quite successful, though no doubt a bit fiddling to set up initially, but it was superseded by the pendulum arrangement seen in figure 6.24, which could provide higher speeds. The plateholder S was fixed to the top end of the pendulum A, pivoted at O. A weight B was attached to the lower end to counteract the plate holder so that the pendulum would rest balanced in any position. The pendulum was first moved so that the pin G was caught beneath the hook at the lower right-hand side of the mounting table. A weight E was then hung from the lever D so that when the handle was pulled up to release the hook, the pendulum would accelerate and move the plate holder to the right. It was so arranged that when the sensitive plate came to the point where the image of the mercury was projected upon it, the weight would hit the table, removing the accelerating force, so that the plate would move onwards with constant velocity. The final plate speed was a function of the size of the weight E, and with a 2 kg weight, a final speed of 150 cm s^{-1} was possible. Having reached the other extremity of its swing, the pendulum was caught and held by the latch H holding onto the pin G. If required for the investigation of transient 'one off' phenomena, the lever M was arranged to lift the contact K out of a mercury bath L in order to trigger the phenomenon under observation.

All this was reasonably simple, but, alas, things were not as straight-forward as they seemed. Burch made a very detailed study of the behaviour of the electrometer under variable-voltage conditions [69,70]. We cannot

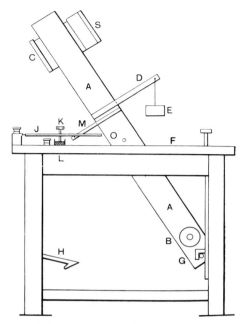

Figure 6.24 'Pendulum motor' which replaced the hydraulic system of figure 6.23 [68] (p436).

possibly discuss all his detailed findings here but, briefly, the complicating feature was that the mercury has mass, and hence inertia, and the response to a voltage step is not instantaneous but gradual as shown in the photograph of figure 6.25. Note that in this trace time runs from right to left and it is also upside down, the mercury being at the bottom. He also showed that the electrometer behaved very much like a capacitor, retaining its charge and deflection for some time after the voltage causing it had been removed. In fact, he discovered that the applied voltage was effectively 'used up' in two ways, one part being proportional to the deflection, the other part to the velocity with which the interface was moving. Thus, given the photographic trace, the determination of the applied voltage involved the measurement of the deflection, and also of the velocity of movement at that time. Of course the photographic plate on the pendulum moved in the arc of a circle, but it turned out that this was rather more convenient than a simple Cartesian plot would have been. In figure 6.26, the rectangle represents the photographic plate. Z is the trace recorded on it as it moves around its circular path and AB is a track recorded at constant radius for ease of measurement. The first component of the applied voltage, that proportional to the displacement Δr, can be found by measuring the radial distances between the two tracks as, for example, at the point p. The velocity at which the interface is moving is obviously proportional to the slope of the trace—i.e. the tangent to the curve. It can be shown that the so-called 'polar subnormal' $o - n$ is directly proportional to $dr/d\theta$, the required velocity. The second component of the applied voltage can thus be obtained by direct measurement of $o - n$. We need not here concern ourselves with the determination of the constants of proportionality of the system; we may simply note that at any time the applied voltage is given by $v = kr + L(o - n)$, where k and L are the aforementioned constants.

Figure 6.25 Trace produced when a voltage step was applied to Burch's oscillograph. Note that time runs from right to left, and the photograph appears inverted, the mercury shadow being at the bottom [68] (p515).

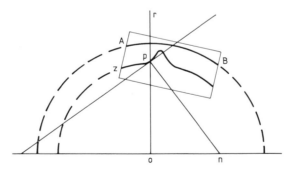

Figure 6.26 Diagram explaining the method of taking measurements from the photographic plate exposed on the pendulum apparatus [68] (p473).

Figure 6.27 is an example published by Burch and shows part of the waveform of the trilled letter 'r'. It was clearly a rather complicated matter to extract from photographic records of this sort the information required to plot the original voltage waveform and the special plotting table shown

Figure 6.27 Trace produced by the trilled letter 'r' spoken into a microphone [68] (p517).

in figure 6.28 was constructed to aid in this work. The photographic plate is first clamped upon the carrier B which is pivoted at O and its precise position is adjusted carefully by means of the screws C and D so that it rotates in the same way, and with the same radius, as it did when it was exposed on the pendulum. The thread F attached to the pivot and kept taut by a small weight is used to align the instrument. The image of the horizontal wire G is cast upon the underside of the plate by light reflected from the inclined mirror through the lens H. The thread lies in a small notch near letter F. The image of the wire is adjusted so that it coincides with the thread, and thereafter the thread can be removed. A glass scale is provided so that the radial movement of the trace can be measured as the plate is moved around the pivot.

A measuring block L which is shown in detail in figure 6.28(c) is then placed upon the photographic plate. This block has a hole drilled through

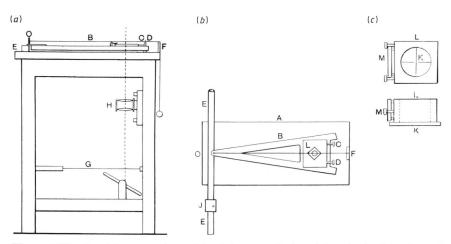

Figure 6.28 Burch's measuring table: (*a*), general view (elevation); (*b*), plan of table-top; (*c*), detailed plan and elevation of the mirror block L [68] (p515).

it, and visible through this hole are two lines at $90°$ which are inscribed on the glass plate K which forms the base of the block. The continuous line is set to be tangential to the curve at the point being measured. On the side of the block is a mirror M, and on its surface is a pointer. Referring now to the plan view of figure 6.28(b); the slider J moves along the rod EE. This slider carries a vertical pin, and it is moved along the rod until the pin, its image in the mirror and the pointer are all in alignment when viewed from the left of the apparatus. The distance OJ measured on a scale, is then the 'polar subnormal' referred to above. If the photograph of figure 6.27 is examined in detail, it will be seen that there are in fact three traces projected onto the plate. The lower trace is the waveform of the sound 'r'. The central line is a fixed trace used as a reference from which the movement of the lower trace is measured. The upper trace would normally have had super-imposed upon it the vibrations of a tuning fork, thus serving as a time calibration for the measurement. In this particular case, the 'r' sound was being spoken into a microphone situated near the apparatus, and so to avoid acoustically generated interference with the main trace it had been switched off. However it can be inferred from Burch's remarks that the plate had been moving at a velocity somewhere between 2 and 3 m s^{-1} and that the total time represented on the photograph is of the order $\frac{1}{3}$ s.

The method of measurement was to determine and tabulate the radial movement and the polar subnormal at each required instant on the waveform and then convert these readings into voltage values by use of the appropriate constants. This was obviously a method involving considerable time and labour and in its development Burch, whose own particular interest was in the electrophysiology of nerves and muscles, had obviously had to expend much effort and ingenuity in overcoming its various com-plicating features. In his own words

> It has qualities of its own distinct from any other form of electrometer or galvanometer, and these will make it a valuable addition to the most perfectly appointed laboratory. I have described the method by which these special properties may be utilised, and some of the apparatus I have designed for the purpose during the nine years I have worked with the instrument, in order if possible to bring it into general use. For the same reason I have refrained from patenting any part of it, preferring to present my discovery and invention to the scientific public.

He claimed that one of its chief merits was that of cheapness—'It is essentially the instrument for those who have to think twice before spending a shilling'. One wonders whether this comment really takes into account the cost of the elaborate measuring apparatus required. Certainly it is another excellent illustration of the lengths to which people were prepared to go, the patient and meticulous attention to detail which would be given, in order to exploit any phenomenon which seemed likely to be useful in the all-important task of registering electrical waveforms.

References

[1] Gerard E M 1912 *Mesures Électriques* 4th edn (Paris: Gauthier-Villars) p633
[2] Bain A 1846 *GB Patent Specification* 11,480
[3] Preece W H and Sivewright J 1914 *Telegraphy* (London: Longmans, Green) pp140–1
[4] Marland E A 1964 *Early Electrical Communication* (London: Abelard–Schuman) p121
[5] Hess A 1897 *L'Éclairage Électrique* **13** 385–90, 455–8
[6] Preece W H and Sivewright J 1905 *Telegraphy* (London: Longmans, Green) pp77ff
[7] Grützner P 1887 *Pflüger's Archiv für Physiologie* **41** 256–81
[8] Grützner P 1900 *Ann. Phys., Lpz* **1** 738–57
[9] Report 1900 *L'Éclairage Électrique* **24** 37–8
[10] Potter R K, Kopp G A and Green H C 1947 *Visible Speech* ch 2 (New York: Van Nostrand)
[11] Janet P 1894 *C. R. Acad. Sci., Paris* **119** 58–61
[12] Janet P 1894 *C. R. Acad Sci., Paris* **119** 217–18
[13] Janet P 1894 *La Lumière Électrique* **53** 92–4
[14] Janet P 1895 *L'Éclairage Électrique* **2** 241–8
[15] Janet P 1895 *Bull. Soc. Int. des Électriciens* **12** 6–17
[16] Barbillon L 1904 *Manipulations et Études Électrotechniques* (Paris: Dunod) pp194–7
[17] Blondel A 1894 *C. R. Acad. Sci., Paris* **119** 399–402
[18] Blondel A 1894 *L'Éclairage Électrique* **1** 83–4
[19] Janet P *see* [14]
[20] Janet P 1894 *C. R. Acad. Sci., Paris* **118** 862–4, 946
[21] Barbillon L *see* [16]
[22] König W 1899 *Elektrotech. Z.* **20** 415–16
[23] König W 1899 *L'Éclairage Électrique* **20** 435–8
[24] *See* [9]
[25] Brown J 1888 *Phil. Mag.* **26** 502–5
[26] Brown J 1896–7 *Nature* **55** 269–70, 294
[27] König W 1900 *Ann. Phys.* **2** (Series 4) 860–2
[28] Peters J F 1924 *Electrical World* **83** 769–73
[29] Report 1924 *Electrical Review* **94** 770
[30] Murphy E J 1902 *US Patent Specification* 713,479
[31] Report 1902 *Electrical World and Engineer* **40** 914
[32] Duddell W D 1897 *Electrician* **39** 636–8
[33] Crehore A C 1895 *Phys. Rev.* **2** 122–37
[34] Kaye G W C and Laby T H 1941 *Physical and Chemical Constants* 9th edn (London: Longmans, Green) p91
[35] D'Arsonval J A 1884 *La Lumière Électrique* **12** 156–7
[36] Bequerel H 1884 *La Lumière Électrique* **12** 321–3
[37] Blondin J 1895 *L'Éclairage Électrique* **2** 337–41
[38] Duddell W D *see* [32]
[39] Pionchon J 1895 *C. R. Acad. Sci., Paris* **120** 872–4
[40] Pionchon J 1895 *L'Éclairage Électrique* **3** 232–3

[41] Abraham H 1897 *Bull. Soc. Int. des Électriciens* **14** 397–434
[42] Abraham H and Buisson H 1897 *L'Éclairage Électrique* **12** 221–2
[43] Switzer J A 1898 *Phys. Rev.* **7** 83–92
[44] Crehore A C and Squier G O 1895–6 *Phys. Rev.* **3** 63–70
[45] Nichols E L 1894 *Proc. Am. Soc. for Advancement of Science, 42nd Meeting, (Madison, Wisconsin) 1893* (Salem: Am. Soc. for Advancement of Science) pp55–71
[46] Feldmann C P 1894 *Wirkungsweise, Prüfung und Berechung der Wechselstrom–Transformatoren* (Leipzig: Oskar Leiner) pp395–6
[47] Duddell W D *see* [32]
[48] Blondel A 1894 *La Lumière Électrique* **51** 172–5
[49] MacGregor-Morris J T and Mines R 1925 *J. IEE* **63** 1056–107
[50] Taylor Jones E 1907 *Phil. Mag.* **14** 238–54
[51] Report 1922 *Engineer* **133** 76–7
[52] Ho H and Koto S 1913 *Proc. Phys. Soc.* **26** 16–27
[53] Bolton D J 1923 *Electrical Measuring Instruments and Supply Meters* (London: Chapman and Hall) p183
[54] Laws F A 1938 *Electrical Measurements* (New York: McGraw-Hill) pp665–7
[55] Golding E W 1935 *Electrical Measurements and Measuring Instruments* 2nd edn (London: Sir Isaac Pitman) p533
[56] Irwin J T 1907 *J. IEE* **39** 617–47
[57] Irwin J T 1925 *Oscillographs* (London: Sir Isaac Pitman) pp42ff
[58] Irwin J T 1907 *Electrician* **59** 266–7, 306–7
[59] Stubbings G W 1929/30 *Commercial AC Measurements* (London: Chapman and Hall) pp304–5
[60] Report 1884 *Electrician* **13** 431–2
[61] Mayençon M 1892 *L'Éclairage Électrique* **45** 627–8
[62] Simmons H H 1912 *Electrical Engineering* (London: Cassell) pp274ff
[63] Lippmann G 1894 *J. de Physique Théorique et Appliqué* **3** 41
[64] Lippmann G 1877 *Phil. Mag.* (Series 5) **4** 239
[65] Lippmann G 1877 *Ann. de Chimie et de Physique* (Series 5) **12** 265
[66] Hospitalier E 1884 *Formulaire Practique de l'Électricien* (Deuxieme anée Masson, L'Academie de Médecine Paris) 97
[67] Waller A D 1887 *Phil. Trans. R. Soc.* B **178** 215–55, 1889 B **180** 169–94
[68] Burch G J 1896 *Electrician* **37** 380–2, 401–3, 435–7, 472–3, 512–17, 532–5
[69] Burch G J 1892 *Phil. Trans. R. Soc.* A **183** 81–105
[70] Report 1892 *L'Éclairage Électrique* **44** 145–6

7

Cathode Rays

Summary

This chapter provides a brief account of the development of the cathode ray method of waveform display from its inception in 1897 to the present day.

The cathode ray oscillograph (CRO), or oscilloscope as it is now more usually known, has become the universally employed method of displaying waveforms, supplemented in recent years by digital storage techniques which enable a permanent record of the traces, ('hard copy' in the current jargon), to be produced on an $X-Y$ plotter. The method has been subject to a continuous process of development for a period of some eighty years, and the history of that development has been well documented [1–3]. The very detailed paper written by Professor J T MacGregor-Morris and R Mines in the *Journal of the Institution of Electrical Engineers* in 1925, and MacGregor-Morris and J A Henley's subsequent book deserve special mention here, as they are excellent sources of reference to the original papers describing the various innovations that have been made. An adequate account of the history of such a successful instrument would really demand a complete book in its own right. The purpose of this concluding chapter, therefore, is not to attempt such a complete record, but rather to remind the user of the modern CRO that it was not always the convenient, ergonomically designed piece of apparatus which we find in our laboratories today.

7.1 The first steps

The gas discharge glow tubes manufactured by Heinrich Geissler in Bonn were well known to the audiences at popular scientific lectures in the second half of the nineteenth century [4,5]. These tubes, twisted into various interesting shapes, and often filled with fluorescent substances to enhance the effects, were improvements on the original simple glass bulb with two electrodes which was sometimes known by the interesting name of 'the electric egg'. Volume III of the book *An Elementary Treatise on Natural*

Philosophy by A Privat Deschanel contains a rather delightful coloured frontispiece which illustrates the various decorative effects which could be achieved using these tubes [6].

In the forty year period 1858–98, many people investigated the mysterious 'cathode rays' which caused the glass and the various other substances in the tubes to fluoresce. Plücker, Hittorf, Goldstein, Wiedemann and Lenard in Germany, Puluj in Austria, and Crookes in Britain, amongst others, all contributed to these investigations, and Crookes' 'Maltese cross' tube was a well known demonstration that the cathode rays travel in straight lines and can cast the shadow of a solid object [7].

It was Ferdinand Braun who, in 1897–8, first adapted the phenomenon to the display of electrical waveforms [8,9]. His tube, shown in figure 7.1, which was constructed for him by Franz Müller who had taken over Geissler's business, consisted of a cold cathode K, and an anode A. The stream of cathode rays, or electrons, was formed into a cylindrical beam (or jet as it was then known) by means of a circular diaphragm C with a small hole at its centre. This jet was allowed to fall on the fluorescent target D, which consisted of a circle of mica, coated with the mineral Willemite (zinc silicate) on its inner surface. A spot of light was created at the point of impact of the jet. In Braun's original experiment, reported in 1897, he placed a solenoid carrying current from the alternating mains supply alongside the tube. The alternating magnetic field caused the spot to oscillate, producing a line of illumination on the screen. When this was viewed through the usual rotating mirror, the waveform of the electrical supply provided by the central Strasbourg power station was revealed. His paper commented on the sinusoidal nature of the supply voltage and compared it with the waveform produced by a vibrating tuning fork, although it is not made clear whether the tuning fork waveform was also made visible by means of the cathode ray tube, or by some other independent method. On August 24th 1897, Dr Braun demonstrated his apparatus to the meeting of the British Association for the Advancement of Science which was meeting that year in Toronto [10].

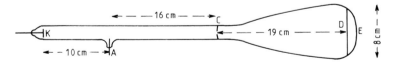

Figure 7.1 Ferdinand Braun's cathode ray tube of 1897 [8].

Shortly afterwards, a certain Monsieur Albert Hess pointed out that he himself had suggested the use of cathode rays for this purpose in a paper written some three years earlier in the French journal *Comptes Rendus*, this having also been reported in other journals such as *La Lumière Électrique*

[11,12]. Actually, Braun's own paper had mentioned in a footnote that 'this application of cathode-rays had been reported by our contributor Monsieur Hess a few years ago'. The footnote actually refers to 'notre collaborateur M. Hess', which might be taken to imply that he and Hess had been close colleagues. However, the French word 'collaborateur' can be variously translated as 'collaborator' or as 'contributor' (to a journal). There is no evidence to suggest any close connection between the two men, and in their biography of Ferdinand Braun, Kurylo and Suskind give an account of the Hess versus Braun episode which leaves no room for doubt on this score [13]. Hess's proposal had been to allow the stream of cathode rays to emerge from a Geissler tube through a thin slit in an aluminium foil window, and to permit them to fall onto a photographic plate situated in an outer evacuated chamber. His paper concluded with the words ' . . . one possesses in the cathode rays a pointer without inertia, capable of recording, with a speed limited solely by the sensitivity of the photographic film, the most rapid variations in the intensities of magnetic fields and, indirectly, the intensities of electric currents'.

It was undoubtedly true that Hess had suggested the use of cathode rays for this purpose several years before Braun and for this reason, many writers of the day (especially the French!) studiously referred to it as the Hess–Braun method [14–16]. However, it was also apparent that Hess had not developed his suggestion; neither did he appear to have actually published any curves recorded by this means. Furthermore, he had not conceived the very practical idea of using a fluorescent screen to reveal the deflection of the rays but had relied instead on the much less convenient method of direct impact on a photographic plate situated inside the evacuated system. His name was gradually dropped, and the device became known universally as the Braun tube. In some German technical dictionaries even today the term 'die Braunsche Röhrer' is given as an alternative to the normal 'die Kathodenstrahlröhrer'.

7.2 Improvements

It must not be thought that Braun's method was immediately hailed by everyone as the perfect answer to the waveform display problem. It was still, after all, an extremely crude instrument which would require many years of patient development to bring it to perfection. Professor André Blondel wrote of it in 1900, [17]:

> The most imperfect method practically used so far is the method of oscilloradiographie (which was what he called the c.r.t.), because of the very complicated materials it requires, the lack of precision of the trace, the feeble sensitivity, the difficulty of operation etc. From these various points of view it seems to be difficult to perfect it, unless one can replace the cathode-rays

with those from radium or a very active body. It also has the disadvantage of introducing a certain inductance into the circuit. But it can nevertheless be called upon to give useful service to study high-frequency oscillations, for the deflection of the cathode rays seems to be virtually instantaneous and without any inertia. This point needs to be fully verified however.

Here the inductance mentioned refers to the magnetic deflection coils.

Many improvements on the simple tube were needed to turn Braun's basic idea into a practical working proposition. Braun's student, and eventual colleague, Johnathan Zenneck improved the electrode system by changing the anode into the form of a pierced disc instead of the side anode of the original tube, and also by using several successive discs to further concentrate the beam. Others, notably Ryan and Rankin in the USA, used a solenoid wound around the neck of the tube in order to improve the focus [18]. The introduction by Wehnelt of the oxide-coated hot cathode in 1904, in order to improve the supply of electrons, was also a considerable step forward, although cold cathodes continued to be used for many years, particularly in tubes with very high anode voltages [19–22]. The early tubes used a very soft vacuum, giving rise to the production of ionised gas particles. This helped to focus the beam, but bombardment of the cathode by positive ions had the effect of shortening its life, and the gradual introduction of hard-vacuum tubes, which relied on the production of a stream of electrons alone (rather than electrons and negative ions) was a step forward. Deflection of the beam by electrostatic means was also tried, and Roschansky, in 1911, constructed a tube in which two pairs of *X*- and *Y*-deflector plates were mounted inside the tube in the manner which is familiar to us today [23]. Glassware manufacturing techniques also needed improvement if larger sizes of tube, properly sealed, were to be constructed [24].

As previously mentioned, the early cathode ray tube was seen as just a natural development of other deflecting instruments such as the loop oscillograph, and so they were used with the familiar methods of time scan such as the rotating mirror [25,26]. Figure 7.2 shows an interesting form of synchronously rotating mirror which was used by A Weinhold in 1901 [27]. The mirror was fixed on a rotor having numerous iron bars mounted around its periphery, and these, in conjunction with the coils e, formed a simple reluctance motor which ensured that the mirror was rotated at a speed locked to the frequency of the mains voltage under observation. Other experimenters employed the well known traditional photographic recording techniques using falling or sliding plates [28–30], or film wrapped around a recording drum [31]. Figure 7.3, for example shows the arrangement used by Wehnelt and Donnath in which the plate holder K is moved manually along the slide V at the back of the camera. A pierced metal disc (external to the tube) mounted on the tine of a tuning fork is illuminated by a light L so as to record simultaneously a time calibration trace.

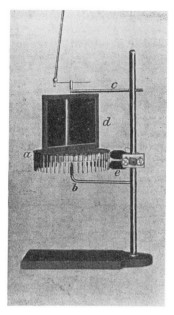

Figure 7.2   Weinhold's synchronously rotating mirror [22]. (The rod at the top is part of the arrangement for bringing the motor up to speed manually.) Photograph: Science Museum (390/86).

One of the most popular forms of display was that using simultaneous deflection in the *X* and *Y* directions, producing a Lissajous type of display which, as previously mentioned, was often known as a 'cyclogram' or a 'cyclographic display'. This was frequently used to demonstrate the phase difference produced by an inductor introduced into an AC circuit. Braun's original 1897 paper described just such an arrangement which he used to illustrate the effect of gradually introducing an iron core into one of the deflecting solenoids. The resulting trace was more or less elliptical in shape, depending on the degree of insertion of the core. A natural extension of this idea was to use a waveform of known shape (usually a sinusoid) for deflecting in one direction, so that by careful measurement the shape of another unknown wave, causing deflection in the other direction, could be determined from the loop [32–37]. The reference sine wave was often derived from the supply mains voltage, any harmonics present being removed by a simple low-pass filter arrangement. Sometimes an extra phaseshift was deliberately introduced into the known sinusoidal waveform so that, to quote one experimenter, 'one can cause the more intricate portions of the unknown wave to occur at the time when the rate of change of the sinewave is most rapid, so that the details are brought out to the fullest extent' [38,39]. Such measurements were, naturally, very time consuming, involving tracing the loop, either manually or by photography, then measuring with compasses and ruler. The Lissajous loop method was also a very good way of determining accurately the frequency of the supply voltage by comparison with a sinusoid of known frequency—often by checking it in some way against a tuning fork [40].

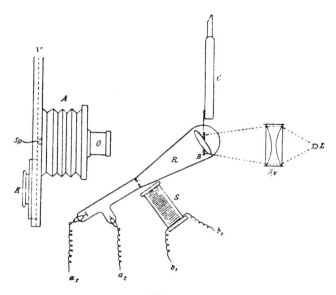

Fig. *a*.

A, appareil photographique : R, tube de Braun ; B, disque d'aluminium, normal à l'axe de l'appareil, dans le plan vertical passant par la tache lumineuse de l'écran, ce disque est fixé au diapason électromagnétique C : Sy, collecteur projetant sur le disque B la lumière de la lampe L : cette lumière converge sur l'objectif O ; a_1, a_2, fils reliant le tube de Braun à la machine électrique ; S, bobine déviatrice, reliée par les fils b_1 et b_2, au circuit du courant étudié : V, planche à glissières, le long de laquelle on peut déplacer le châssis K.

Figure 7.3   Sliding photographic plate used with cathode ray tube [29].

It was obvious from an early stage that it would be highly advantageous to impose a linear motion on the beam in one direction so as to produce a true voltage–time trace. In 1899, Zenneck produced just such a display by use of the piece of apparatus shown in figure 7.4 [41–43]. It consisted of a rotating wheel having a conducting wire fixed around its periphery, with two slip rings to enable the ends of the wire to be connected to the terminals of a battery. The beam-deflecting coil was connected between one end of the battery and a stationary brush bearing on the edge of the wheel. The arrangement was precisely like a modern rotary potentiometer, save that the slide arm was fixed, and the main body of the potentiometer was rotating. In this way, a current having a linear sawtooth form was caused to flow in the solenoid, which thus produced a linear scan of the spot across the screen as on a modern cathode ray oscillograph instrument. The circular or polar timebase also had its adherents, and the early history of this is recounted in a paper written by D W Dye in 1925 [44].

J A Fleming, amongst others, used a rather different approach. He scanned in the X-direction using a sinusoidal current, but made the extent of the scan so large that during the time the spot was actually moving across the comparatively small fluorescent screen its motion was virtually linear. He used this to investigate the discharge of capacitive circuits [45,46]. It was Rogowski, in the early 1920s, who introduced what we would recognise as a proper timebase generator using the linear charge and discharge of a capacitor through saturated diodes to produce an almost linear electrostatic scan, although he did not at first use the principle of fast flyback, but scanned alternately in each direction across the screen [47]. He subsequently suggested allowing the capacitor to discharge across a spark gap in order to produce rapid motion of the beam in the backward direction. In 1924, N T Kipping introduced the neon timebase in which discharge of the capacitor through a neon lamp produced the familiar sawtooth scan [48].

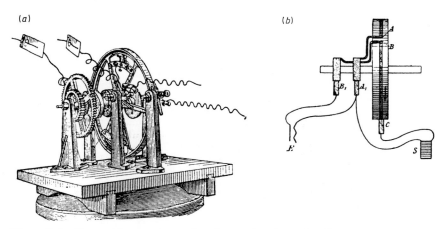

Figure 7.4 Zenneck's apparatus for the production of a linear sawtooth scan: (*a*), general view; (*b*), detailed view of conducting track showing connection to deflection coil S [43].

In the 1910s and 1920s there was also a considerable vogue for the use of large cathode ray tubes with cold cathodes and very high accelerating voltages, one of the chief advocates of these being Professor Alexandre Dufour of the Sorbonne University [49–54]. In tubes of this sort, the movement of the electron beam was recorded on a photographic plate or revolving drum situated inside the evacuated tube. This involved continual opening up and evacuation of the tube, and cannot have been very convenient, to say the least. Anyone who has ever complained about the complexity of operation of the controls on a modern oscilloscope should perhaps contemplate figure 7.5, which shows an actual photograph of Dufour's apparatus. For recording traces of low-frequency signals, he used a conventional revolving drum, arranging an additional slow deflection of the beam so as to produce a spiral trace. For high-frequency work, however, he used a stationary plate, and by means of subsidiary deflecting signals produced a zig-zag, or side-to-side trace of the type shown in figure 7.6. This particular example shows the rather fuzzy record of a train of damped oscillations of frequency 8.5 MHz. Dufour's main aim was to adapt the cathode ray oscillograph to the recording of transient, or one-off

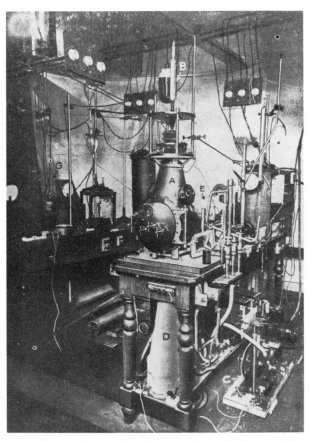

Figure 7.5 Dufour's oscillograph [50] (pp638–63). Photograph: Science Museum (398/86).

phenomena [55]. One of the problems with this type of tube was fogging of the film due to the impact of stray electrons whilst waiting for the transient to occur. This was sometimes overcome by the incorporation into the tube of what was called a Norinder relay, which was a deflecting device which interrupted the beam by directing it on to a target during the waiting period, allowing it to proceed into the photographic chamber only during the brief period of the occurrence of the transient phenomenon [56,57].

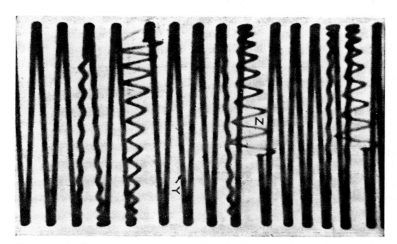

Figure 7.6 One type of trace used by Dufour for the study of high-frequency signals (8.5 MHz damped oscillations) [50] (pp638–63).

Development of tubes using lower voltages was also proceeding, and by the 1930s commercial instruments were becoming available. MacGregor-Morris and Henley in their book, previously mentioned, published in 1936 give details of many of the commercial tubes then being manufactured, both gas-filled to focus the beam, and hard-vacuum types [58]. Some of the complete instruments look rather odd to the modern eye, since they came in two parts connected by a length of cable; one part contained the tube, the other part the associated power supplies and other circuitry [59,60]. The magazine *The Engineer* of 1936 contains an interesting article describing a portable cathode ray oscilloscope housed in a caravan, which had apparatus which could photograph simultaneously the traces from up to twelve tubes. This was intended for the observation of transient phenomena on transmission lines 'in the field' [61,62].

It was, of course, the Second World War and the development of Radar which gave a tremendous boost to electronic techniques in general, and the development of the cathode ray tube in particular. By the time the war was over, the CRO was fully established as an indispensable laboratory tool, and the well known Cossor double-beam oscilloscopes, (available in AC- or DC-coupled form and optionally equipped with long-persistence phosphor screens to assist in the observation of slow or transient phenomena), were the familiar work horses of British laboratories. These were provided with a ruled graticule in front of the fluorescent screen, but time and voltage measurements had to be performed with the aid of calibrated shift controls, a system which was neither very convenient nor very accurate. The 1950s saw the spread of instruments, notably the Tektronix range, in which the graticule was actually calibrated in $V \, cm^{-1}$ and time cm^{-1}, which was another great step forward. Phosphor storage tubes also became available, which allowed the trace to be retained on the screen for many minutes. These were invaluable for the study of transient phenomena and for quasiperiodic signals such as speech. Oscilloscopes incorporating tubes of this sort are still available today, but in recent years they have been supplanted for the most part by digital storage types. The modern trend is to provide instruments which not only perform the duties of a normal oscilloscope, but which also incorporate digital computer circuitry which will produce directly on the screen such computed functions as amplitude spectra, correlation functions, power spectra etc., almost literally in the proverbial 'twinkling of an eye'.

The science of 'waveform tracing' has indeed come a long way since the days of Wheatstone's mirror, König's flames and Léon Scott's phonautograph.

References

[1]   Parr G and Davie O H 1959 *The Cathode Ray Tube* 3rd edn (London: Chapman and Hall)
[2]   MacGregor-Morris J T and Mines R 1925 *J. IEE* **63** 1056–107
[3]   MacGregor-Morris J T and Henley J A 1936 *Cathode Ray Oscillography* (London: Chapman and Hall)
[4]   Plücker J 1858 *Ann. Phys., Lpz* **103** 88–106
[5]   Plücker J 1858 *Phil. Mag.* (Series 4) **16** 119–35
[6]   Deschanel A P 1881 *Elementary Treatise on Natural Philosophy: Part III Electricity and Magnetism* 6th edn (Translated by J D Everett) (Glasgow: Blackie)
[7]   Crookes W 1879 *Phil. Trans. R. Soc.* **170** 135–64
[8]   Braun F 1897 *L'Éclairage Électrique* **12** 131–2
[9]   Braun F 1897 *Wied. Ann.* **60** 552–9
[10] Braun F 1897 *Report of British Association for the Advancement of Science* (*Toronto*) *1897* p570
[11] Hess A 1894 *La Lumière Électrique* **53** 91–2
[12] Hess A 1894 *C. R. Acad. Sci., Paris* **119** 57–8
[13] Kurylo F and Suskind C 1981 *Ferdinand Braun* (Cambridge, MA: MIT) p98
[14] Anon 1900 *Nature* **63** 142–5.
[15] Abraham H 1897 *Bull. Soc. Int. Elec. Eng.* **14** 397–434
[16] Vigneron E 1907 *Mesures Électriques* (Paris: Gauthier-Villars/Masson) p62

[17] Blondel A 1900 *Congrès Int. de Physique* (*Paris*) *1900* **3** 264–95
[18] Rankin K 1906 *US Patent Specification* 838,273
[19] Ackerman O 1930 *Instruments* **30** 775–7
[20] Anon 1934 *Engineer* **157** 36–8
[21] Ackerman O 1930 *Trans. Am. AIEE* **49** 467–75, 1930 285–9
[22] Anon 1930 *Instruments* **3** 786–90
[23] Roschansky D 1911 *Ann. Phys., Lpz* **36** 281–307
[24] Ryan H J 1903 *Trans. Am. IEE* **22** 538–552
[25] Simon H T and Reich M 1901 *Phys. Z.* **2** 284–91
[26] Mayrhofer G 1900 *Elektrotech. Z.* **21** 913–15, 926–9
[27] Weinhold A 1901 *Elektrotech. Z.* **22** 409–11
[28] Wehnelt A and Donath B 1899 *Wied. Ann.* **69** 861–70
[29] Wehnelt A and Donath B 1900 *L'Éclairage Électrique* **23** 230–2
[30] Wehnelt A and Donath B 1900 *Elektrotech. Z.* **21** 103
[31] Oosting H J 1900 *Phys. Z.* **1** 177–9
[32] Ryan H J *see* [24]
[33] Rankin R 1905 *The Electric* (*Club*) *Journal* **2** 620–31
[34] Ryan H J 1906 *US Patent Specification* 834,998
[35] Ryan H J 1903 *Trans. Am. IEE* **20** 1417–30
[36] Seefehlner E E 1899 *Elektrotech. Z.* **20** 120–1
[37] Petrovski A 1904 *Electrical World and Engineer* **43** 812–14
[38] Ryan H J 1903 *Electrical World and Engineer* **42** 103
[39] Ryan H J 1903 *Electrician* **51** 770–2
[40] Zenneck J 1899 *Wied. Ann.* **68** 365–8
[41] Zenneck J 1900 *Elektrotech. Z.* **21** 102
[42] Zenneck J 1899 *Wied. Ann.* **69** 838–53
[43] Zenneck J 1900 *L'Éclairage Électrique* **23** 228–30
[44] Dye D W 1925 *Proc. Phys. Soc.* **37** 158–68
[45] Fleming J A 1912 *Proc. Phys. Soc.* **25** 227–9
[46] Simon H T 1905 *The Electrical Engineer* **36** 561–4
[47] Rogowski W 1920 *Archiv. für Elektrotechnik* **9** 115
[48] Kipping N V 1924 *Wireless World* **13** 705
[49] Dufour A 1923 *Oscillographe Cathodique* (Paris: Chiron)
[50] Dufour A 1922 *L'Onde Electrique* **1** 638–63, 1923 **2** 19–42
[51] Dufour A 1924 *C. R. Acad. Sci., Paris* **178** 1478–80
[52] Wood A B 1925 *J. IEE* **63** 1046–55
[53] Wilson W 1943 *The Cathode Ray Oscillograph in Industry* (London: Chapman and Hall)
[54] Laws F A 1938 *Electrical Measurements* (New York: McGraw-Hill) pp675ff
[55] Dufour A 1914 *C. R. Acad. Sci., Paris* **158** 1339–41
[56] Anon 1930 *Instruments* **3** 786–90
[57] Ackerman O *see* [21]
[58] MacGregor-Morris J T and Henley J A *see* [3]
[59] Anon 1933 *Instruments* **6** 19, 178
[60] Turner H C and Banner E H W 1935 *Electrical Measurements* (London: Chapman and Hall) p123
[61] Anon 1936 *Engineer* **161** 41–4
[62] Wilson W *see* [53], p44

Index